Kinari Webb

From one guardian
to
another!

KWebb

www.guardiansofthetrees.org

GUARDIANS

of the

TREES

GUARDIANS
of the
TREES

A JOURNEY OF HOPE THROUGH
HEALING THE PLANET

Kinari Webb, M.D.
Founder of Health In Harmony

FLATIRON
BOOKS
NEW YORK

www.flatironbooks.com

Map originally designed by Grapheon Communications Design and modified
by Health In Harmony for publication in this book

Library of Congress Cataloging-in-Publication Data

Names: Webb, Kinari, author.
Title: Guardians of the trees : a journey of hope through healing the planet / Kinari Webb, M.D.
Description: First edition. | New York : Flatiron Books, 2021.
Identifiers: LCCN 2021024209 | ISBN 9781250751386 (hardcover) | ISBN 9781250751409 (ebook)
Subjects: LCSH: Webb, Kinari—Health. | Environmental health. | Forests and forestry—Health
    aspects. | Rain forests. | Human ecology.
Classification: LCC RA565 .W43 2021 | DDC 613/.1—dc23
LC record available at https://lccn.loc.gov/2021024209

Our books may be purchased in bulk for promotional, educational,
or business use. Please contact your local bookseller or the
Macmillan Corporate and Premium Sales Department at 1-800-221-7945,
extension 5442, or by email at MacmillanSpecialMarkets@macmillan.com.

First Edition: 2021

10 9 8 7 6 5 4 3 2 1

*to my child, who is such an unexpected blessing in my later years
and is being born just as this book is*

A Note on Names, Translations, and Events

Throughout this book, with a few exceptions I use the traditional Indonesian method of addressing people: an honorific plus their first name. Most of the honorifics used are family terms, which are applied to the entire community as a traditional courtesy. For example, any man who has had children will be called *Pak,* which means "father," and similarly, a woman will be called *Ibu,* meaning "mother." A woman who does not have children is called *Tante,* which means "Auntie," and this was how I was often referred to outside the clinic before I had a child. In the clinic and in official settings, I am called Dr. Kinari (or, affectionately, Dr. K). A young person is addressed using sibling terms that indicate their relative age: *Mbak* or *Kak,* "older sister"; *Bang,* "older brother"; and *Adik,* "younger sibling." In Indonesia, in guessing someone's age, you should guess on the *older* side, because it honors them more—even though people may collapse in laughter if you get it wrong. Only if you are very, very close to someone can you use their name without any honorific at all. So I have not used an honorific for people I consider to be my close friend.

In quoting a conversation with an Indonesian, I have generally given only the English translation, even though (unless otherwise noted) those conversations were in Indonesian. (The translations are my own and are sometimes slightly loose so that the meaning is clear.) I have used people's real names where appropriate, but I have changed the names of patients unless they gave me specific permission to share their stories. In some cases, I have changed the person's gender or other details to preserve their anonymity. In one case, I combined two meetings for clarity, but all other events are described as they occurred.

When you live in the forest, it's easy
to see that everything's connected.
—Dr. Jane Goodall

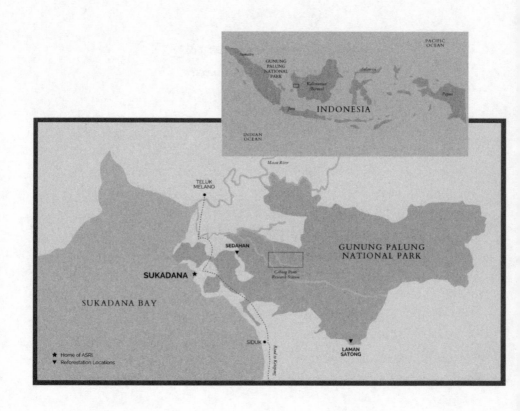

Preface

Growing up in northern New Mexico, I shared a horse, Pinto, with one of our neighbors. One day, Pinto and I were cantering along the mesa top when he saw something. I don't know what it was, but he reared straight up and pivoted around. If you have ever heard the term *runaway horse*—it is real. Pinto laid his ears back and full-out galloped straight for the edge of the mesa. I knew for certain that we would both die if we went over the edge. But Pinto was not responding to the reins, no matter what I tried. If I jumped off, neither of us would survive. Seconds from disaster, I remembered something I once read—that it is impossible for a horse to run with something hanging from its neck.

So . . . I swung both feet over to the same stirrup, grabbed his mane with my right hand, and swung myself down under his neck at full gallop, gripping his mane on both sides. Pinto slid to a stop, just feet from the edge of the cliff, saving both our lives.

THAT DAY WITH PINTO TAUGHT ME an important lesson. Even when things look hopeless, it might still be possible to avoid disaster. This book is about my own work on what feels like the biggest cliff any of us could possibly face: the danger that our earth will no longer be livable for us. Global warming is happening faster than even the worst models predicted ten years ago. We are only beginning to understand just how dangerous the feedback loops can be, as reflective ice disappears, polar seas absorb sunlight and warm more quickly, and melting tundra releases methane into the atmosphere. The Intergovernmental Panel on Climate Change has said that if we don't halve carbon emissions by 2030, we may go beyond irreversible tipping points, at which time it won't matter what we do. In other words, that cliff is very close.

When most people think about emissions, they usually think only about burning fossil fuels for cars, airplanes, and electricity. They don't realize that deforestation releases as much carbon as the entire transportation sector for the whole world. Despite covering only 2 percent of the surface of the earth, rainforests support *50 percent* of the world's species and hold 25 percent of the world's carbon, stored in both the trees and the soil. In addition, as trees grow, they continue to suck down carbon from the atmosphere.

Trees, and especially tropical rainforest trees, are our absolute best friends in halting the climate crisis, since they absorb a third of the pollution we emit every year. They are the natural climate solution, exquisitely invented through millions of years of evolution. Rainforests are huge banks of carbon, and the bigger the tree, the more carbon it absorbs as it grows—and the more oxygen it releases. Forests are for this reason sometimes referred to as the "lungs of the earth." When forest is logged or burned, most of that carbon gets released into the atmosphere, and the local cooling effects of the trees are lost as well. We do have to transition our world to one powered by alternative energies, but we also have to protect the tropical rainforests of the world, because even if we stopped releasing more carbon, we would need the forests' help to get the earth's atmosphere back into a safer balance.

No humans, in all of human existence, have lived through a time like this. Can we stop this runaway horse before we tumble off the cliff?

Can we find a way to live sustainably and in balance with our planet? Is it possible for that transformation to be unexpected and profound?

For more than a decade, I have had the privilege to lead a tiny band of people resisting the loss of biodiversity and the poisoning of our earth, serving as a doctor who heals both people and nature. I founded two organizations that support this work, from different sides of the earth: Health In Harmony in the United States, and Alam Sehat Lestari in Indonesia. The Indonesian name translates to "healthy nature ever-lasting," and the short form, ASRI (pronounced "AHHS-ree"), means "harmoniously balanced." I have had the incredible joy of partnering with my Indonesian colleagues as they transcended poor education, lack of opportunity, and a culture of fatalism, to create a multi-award-winning model that honors the local communities as the wisdom holders for how to reverse rainforest loss and improve human well-being.

I hope this story will inspire and motivate you to join in this movement across the planet to bring about change. By listening to rainforest communities and partnering around the world, we can have huge impacts on the health of the whole planet. By working together, we can thrive.

Death and the Sun

On the equator, dusk falls so quickly it often leaves me feeling bereft, though the sunsets and orange glow of the light make up in brilliance what they lack in length. One night in July 2011, my husband, Cam, and I were swimming in the warm waters of the South China Sea, watching the sunset from the special cove where we retreated most evenings. We couldn't see a single dwelling—although it was only about ten minutes' walk along a sparse path to the thatched roof house that had become our home in the little town of Sukadana, in Indonesian Borneo.

Silvered leaf monkeys chattered in the patch of mangrove behind us, competing with the cacophony of forest insects. One peak of the mountain chain of Gunung Palung National Park rose high above the mangroves, shrouded in wisps of mist and appearing deceptively close. The low clouds began to be lit by bits of pink, and a pair of sea eagles called to each other overhead. I found myself unexpectedly content. It was the perfect ending to a long day of caring for patients in the crowded clinic.

I had helped open the clinic four years earlier, as part of a conservation project designed to improve people's health as the key to saving rainforest. This was a dream I had conceived eighteen years prior—about twenty miles from this very spot, deep in the forest where Cam and I first met, when I was studying orangutans on the western side of Indonesian Borneo. Cam had continued his research in ecology, striving to understand the complexities of tree species diversity in these magnificent forests. I, on the other hand, had taken a new direction—to become a physician, one who believed deeply that, unless both the environment and the people are healthy, neither can be. And now I was beginning to see, with great joy, that this ambitious strategy was no longer a fragile and untested idea for me but a solid success. In just four years, the staff had grown from the original eight to more than sixty dedicated individuals—and we were already seeing dramatic drops in illegal logging, in addition to impressive improvements in community health.

As Cam and I paddled and splashed, enjoying that golden sunset, I was suddenly overwhelmed by searing pain. My right hand reacted even before my brain was conscious of what was happening. I grabbed at the rubbery tentacles that coiled around my left arm and down my back, where they curled into the scoop of my bathing suit. When I pulled at the lashes, they stung my hand fiercely, but I didn't let go. Every fiber of my being knew that my life depended on getting them off me. I was scrambling backward toward the shore, screaming, frantic. Neither of us saw the jellyfish, but we both knew that was the only thing it could be.

Reaching the shallow water, I grabbed sand and scrubbed the sting, but that just seemed to make it worse. What were you supposed to do for these toxins? Vinegar? Urine? The excruciating pain felt like hundreds of wasp stings, so anything was worth a try. I looked up at my husband, who had just caught up with me.

"Cam, can you pee on me?" He obligingly whipped it out, while I crouched in the shallow waves, and . . . nothing happened. "Hurry, Cam, I think I'm going to pass out!"

"I'm sorry—it's just not that easy to pee on your wife."

When it finally did happen, it wasn't pleasant. Nor was there any noticeable change in the fierce pain. Staggering out of the water and sloshing myself clean, I gave Cam the few instructions I could think of. He wasn't a medical doctor, but he was one of the most intelligent people I had ever met, and he had seen me through medical school. "If I lose consciousness, carry me to the house and give me a shot with the EpiPen." I told him to call the two doctors I had been training and quickly ticked off the medicines they should give me.

Every step of the walk home, I doubted I would make it, but somehow I got there. We tried squeezing lime juice on the sting. And then I did the one thing you should never do: I went into our little lean-to bathroom and scooped cold water out of the half-rusted fifty-gallon drum and poured it over my body. The water ran off my skin and down through the wide-spaced plank floor and onto the ground below. I began to scrub the line of welts with soap—but before I could even finish, I started shaking from head to foot. I could feel the waves of toxin flowing through my veins as the fresh water triggered the popping of thousands of tiny pneumatocysts—perfect microscopic balloons of poison, each equipped with a tiny syringe injecting death.

My heart subliminally told me that it was about to stop. As a physician, I found this fascinating—*Ah, patients know when their heart is about to fail.* I could barely breathe, choking and moaning, in indescribable pain. I stumbled back into the house, where Cam caught me as I fell to the floor. Working in a place with minimal pain medication and almost nonexistent health services, I have seen pain—lots of pain. But I had never seen anyone acting the way I was acting, except sometimes women in the last pushes of childbirth. But this was different; intense fear was layered on top, because I knew my heart was moments from pumping its last pump.

People in the village had warned us not to swim in the sea because of the *ampai*. We knew the word meant "jellyfish," but we thought they meant the familiar moon jellies, whose small stings were an irritating but minor nuisance. So, stupidly, I had not listened to them and had swum in that bay probably more than a thousand times. Now I got it— they must have meant box jellyfish, whose toxin is the deadliest thing on the planet. The firebombs exploding in every muscle could come from nothing else. No wonder they had warned us.

Cam scrambled through storage containers until he found the syringe he was looking for. He plunged the EpiPen into my thigh so hard I knew it would leave a large bruise—if I survived. The epinephrine flooded into my body. Half a second later, I felt my heart leap. I held my breath as a sense of relief reached my brain; at least the cardiac part of the toxin wouldn't win at that moment. But every nerve in my body told me this battle was only beginning. Cam used his phone to google "box jellyfish" and discovered that their venom is one hundred times stronger than that of the king cobra we had once had in our kitchen. How could I possibly survive this?

Many hours later, it was 2:00 a.m., and at least fifty people were crowded with me in the clinic. Most of us, including myself, were sure I would die. The action of the neurologic part of the toxin was bringing every muscle in my body to contractions that felt like a seven on a scale of ten—*every* muscle. Then, one after another, each went to the maximum contraction of ten: the perfect outline could be traced beneath the skin, where my muscles stood out in the most excruciating cramp one could imagine. Each one was horribly painful when it fully contracted, but the absolute worst—and most dangerous—was the diaphragm. With the diaphragm in spasm, one cannot breathe. And that was what was happening to me. I pursed my lips against the pain, trying desperately to suck in a tiny bit of air. I writhed almost constantly, no position tolerable.

In lucid moments, I wished for an anatomy book. Muscles I only dimly recalled from medical school would announce themselves in a visible outline and leave me crying. When the one that controls the tiniest bone in the body, the stapes bone in the inner ear, sent out stabbing pain, it was all I could do to keep from screaming.

We administered four times the maximum dose of every pain medication we had in our pharmacy. The wonderful young doctors I had trained were too scared to give me more than that, for fear of killing me. I knew they were right, even as I struggled to breathe. I looked at my left arm, welted from the stings. I very seriously considered amputating it. We had the tools to do it. It would have to be done without anesthesia, but given the nine-out-of-ten pain I was already in, ten out of ten felt worth it. I decided not to amputate, though, only because the venom had clearly already spread throughout my whole body.

Evacuation insurance was one of the benefits of being an adjunct faculty member at the Yale School of Medicine—but the nearest runway was two hours away, and it only operated during daylight. Besides, as we learned from an internet search on the clinic's satellite connection, there was no Western medical treatment for this jellyfish venom anyway. The only thing I could think of that a hospital could do would be to paralyze me and put me on a ventilator. That honestly sounded pretty good, but given that every hour had increased the pain, I doubted I would make it to dawn.

As word went out that I was dying, more and more people gathered. First, about twenty staff members arrived, and gradually, around thirty people from the community showed up in the middle of the night. People's attention soon focused on an argument about potential treatment options. The son of a Chinese shopkeeper in town (whose own life we had saved dramatically—twice) was disagreeing with Ibu Ateng, the traditional massage healer to whom we often referred patients.

"The fisherman who was stung last year only got about *this much* of the tentacle on him." Ibu Ateng indicated the first joint of her thumb, about an inch. "But he was in the hospital for four days! He screamed and cried the whole time, in spite of all the pain medication. How is Dr. Kinari going to survive when she got the tentacle all the way up her arm and all the way down her back? Plus, she bathed with fresh water! She should have known not to bathe for a week!"

The shopkeeper's son held his ground. "The fisherman almost died because he wasn't made to vomit. The evilness stayed in him. If we can just get the *dukun* here to make her vomit, she might survive!"

Hearing this exchange, one of our best nurses begged me, with panic in his eyes, to let him call two *dukuns,* traditional healers. "Dr. K, *please,* maybe they can help you!"

I knew that if I died and the staff hadn't tried every possible remedy, they might never forgive themselves. It was also entirely possible that the community might never forgive them. I didn't want to put them in that position. In any case, I had nothing to lose.

The *dukuns* they were talking about come in many different varieties: women who know about herbs, wise religious healers, and crafty-eyed men who know the famous Melayu arts of witchcraft and poisoning. After much consultation, two healers were invited. One was an old woman with kindness in her eyes, wearing a faded batik sarong. She rubbed herbs on the sting—but the painful scrubbing made my muscles contract even more. Then the next *dukun* stepped forward, an old man so tiny that he would not have come up to my shoulder had I been standing (although I am not tall). He had me drink a glass of brilliant green liquid. This concoction of leaves from a rainforest tree did its job of inducing vomiting amazingly well. In fact, I have never seen a better emetic. Within moments, I was vomiting over and over again, under the watchful eyes of the dozens of people crowded into the hot examination room.

"Pak Alvaro vomited up a live fish. I was there and I saw it!"

"I heard that someone vomited sand. And, Ibu Ateng, didn't you vomit green leaves?"

People were clearly disappointed to see only clear fluid. Now the old man said an incantation over a glass of water and urged me to drink that, too. But for some reason, here I drew the line, and the sweet young Javanese doctor, Dr. Ruth, agreed with me. She pulled the curtain closed, shielding us from view, and stowed the charmed glass out of sight behind the EKG machine. Out of ideas, people stood back as I continued to struggle in overwhelming pain, trying to suck in air using every muscle I could call on to help.

I realized I would probably soon die in my own clinic. I had seen at least ten patients die in this very room. When we first opened the clinic, I was overwhelmed by the large extended families that would gather when someone was critically ill, and I would try to shoo people out. But in later years, I always just let them crowd around. I grew to value this outpouring of love and care. Even children would come, and afterward they often touched the corpse with curiosity, or kissed the body.

After the death, the deceased is taken home, and the whole community gathers. Muslim tradition dictates that the body must be buried before sunset on the day of death. The body is ceremonially washed and wrapped in white cloth; a coffin is made from a freshly cut tree; food is cooked; and wooden posts are carved to mark the head and the foot of the grave. Once everything is prepared, a procession accompanies the family to the cemetery, where a hole has already been dug. In goes the body—and I have always been struck by the way the young men jump down onto the coffin and pack the sand with their bare feet, slowly moving higher as more sand is pushed in from above.

I have sat many times at those funerals, and family members always tell me how grateful they are that we did our best, but when Allah calls, there is nothing that can be done. I was so thankful for this wisdom after working in the United States, where death is always someone's fault—and chances are, it is the doctor who is blamed. If every physician were perfect, people seem to think, no patient would ever die, and everyone would live forever. But here in a tiny village in Borneo, people recognize that no one escapes death. Sitting under the frangipani trees that shed their fragrant white flowers around the graves, listening to the chanting and prayers, always deepened my knowing that death comes for us all.

But this was different—the immediacy of death was a whole new experience. As the wise Romanian writer Petru Dumitru Popescu observes, "There are two things you can never look fully in the face—death and the sun."

By 3:00 a.m., I was looking at the sun—or maybe looking at it the way one looks at an eclipse, reflected upon the ground through a pinhole in a leaf, watching a shadow move over my life. Perhaps the wisdom of my friends in this remote corner of the world was seeping into me; if death called, I would simply accept.

I had fought the good fight, striving to save rainforest and improve the lives of people who faced incredible hardships. Now, surprisingly, I found myself ready to go. I had always been painfully aware of my failings—an intense temper, occasional intolerance of people's shortcomings, and self-centeredness. But at the end, somehow, none of that mattered. Even my flaws seemed to have had purpose. The bigger choices I'd made seemed simply right. I even had a private chuckle, remembering the warnings of some good friends who told me I was crazy to work for no salary in a remote village in Indonesia and not save money for retirement. They appeared to be wrong, since, apparently, I wasn't going to make it to retirement! And the years I had had were profoundly meaningful and even fun. I felt a profound sense of grace—deeply grateful for life and for the people I had shared it with. As a friend had said, in the end, we are all as perfect as a grain of sand.

In a dispassionate, rational way, through the crippling pain, I knew I didn't have much more time. Every breath was already a fight, and with every hour, the pain worsened. I found myself completely unafraid. Cam had not left my side for a moment. He was alternately massaging my muscles and supporting my weight as I tried to find a position I could tolerate or one where I might manage to suck in a breath of air. His worried brow and tender care conveyed all that needed to be said.

I began to say my goodbyes. First to Cam: "My darling, I do not regret anything. I love you completely." I needed to say this, because many years of our marriage had been hard, very hard, and six months earlier, I had almost left him. I wanted him to know that I didn't regret our time together or the separation. We had both grown in the process, and we were consistently happy together for the first time since we met. Our difficulties were the fault of both of us and neither of us. We had each accomplished more in those years together, personally and professionally, than we could have done alone. And we had always loved each other. Dr. Ruth and Dr. Verina averted their faces, and I saw tears in Verina's eyes as they gave us a special moment to kiss each other and say goodbye.

Hotlin Ompusunggu, my dear friend and cofounder of the clinic, had been pacing around me all night. As a dentist, she showed her care by folding squares of cloth for me to bite on so I wouldn't break my teeth when the jaw muscles clenched. Now I held her close and gasped, in fragments, "I am so proud of you! You have done such a wonderful job. I know you can run the program on your own! You can do it! I love you."

Etty Rahmawati—her green headscarf framing her intelligent black eyes, now streaming with tears—held me as we cried together, telling each other how grateful we were for each other's friendship. A line of doctors, nurses, conservationists, patients, neighbors, and friends came to say goodbye, clasping my hands, soothing my brow, or dropping light kisses on my hair. Protestants, Muslims, Catholics, and Confucian Buddhists, representing at least ten different Indonesian ethnic groups, all mourned together and comforted one another. This was the most diverse family that had ever gathered at a death in our clinic, and I felt surrounded by love. It was a good way to go.

OBVIOUSLY, I DIDN'T DIE THAT night. I survived thanks in part to the U.S. poison control hotline that we eventually figured out anyone in the world could call. When Cam explained the situation, they said they had no idea, but they promised to find someone who did. A few hours later, they got back to us, having tracked down a doctor in Thailand who had experience treating box jellyfish stings.

The Thai doctor's advice was emphatic: "*More* pain medication!"

His experience was that a doctor will never encounter patients more in need of analgesics. "Give her enough medicines until she can breathe. Ignore the maximum doses!"

So that is what we did, and I will always remember that moment of relief when I finally started getting full breaths of air into my lungs.

But the ordeal was far from over. For the next few months, I exhibited classic symptoms of heart failure—but I basically just ignored them (doctors are terrible patients!) and slowly improved. The following month, in September, I thought I was well enough to do a fundraising tour in the United States, but midway through the trip, after giving a talk to about a hundred Smith College students, I simply collapsed. My heart rate was through the roof, my blood pressure was tanking, and my respiratory rate went completely haywire.

Cam made the fastest trip a human has likely ever done from a remote island off the coast of Ambon, where he was doing ecological research, all the way to my hospital bed at Brigham and Women's Hospital in Boston (including walking out of the forest through the night, hitchhiking, chasing down the ferry by motorcycle when he just missed it at one port, and buying multiple tickets in airports minutes before flights took off). He met me in Boston, where my father had already flown in from D.C., for my two-week admission, which was followed quickly by yet another hospitalization.

No one could figure out what was wrong with me, and every day, my health deteriorated. It turns out that dying surrounded by loved ones, in a remote village clinic in Borneo, was vastly preferable to enduring a hostile mega-hospital, with overworked and exhausted staff who were irritated that I didn't fit into a nice, easy treatment category. One of the team of specialists memorably declared, "Your symptoms can't possibly be from the jellyfish sting because *it isn't in the literature.* We can't assume this is a *zebra* if we haven't ruled out *horses.*"

I raised my eyebrows at him. "You are kidding, right? How much *literature* do you think is coming out of Indonesian Borneo? The few doctors are so busy saving lives, they don't have time to write articles about all the amazing toxic creatures that aren't even known to science. I really was stung by a highly poisonous jellyfish, and all my symptoms started right after the sting. Last year, I climbed Mount Kinabalu, which is over thirteen thousand feet—and now I can't walk across the room! People from the village I live in say that you can be so sick from being stung that you can't get out of bed for four years! Just because no one has published it yet doesn't mean it isn't real."

But after ruling out everything else with intensive tests and daily lunchtime consultations with multiple specialists, they finally accepted that whatever was wrong with me was caused by the jellyfish. At the end of my stay, my internist summed up what they had concluded. "Well, Kinari, all we can say is that your autonomic physiology is profoundly abnormal, we have never seen anything like it, we assume it's from the box jellyfish, and we have no idea what to do about it."

He clearly felt that they had failed, but I was pleased with that assessment and agreed with it. We had tried various medicines, and nothing worked. The fact is, no matter how much Western medicine likes to believe otherwise, we simply don't know everything; and in my experience, there is more mystery and magic in the universe than we like to admit. It was frustrating, though, to realize that the money the insurance company had spent on my care was enough to care for many thousands of patients in Borneo.

For all that money, they learned that the jellyfish had apparently damaged the part of the spinal cord that regulates the unconscious nervous functions that are necessary for life: heart rate, blood vessel tone, breathing rate, blood pressure, and the complicated coordination that is required for standing, sitting, walking, climbing stairs, coughing, and concentrating. Marvelous, amazing, *simple* actions that we only fully appreciate when they don't work. At one point, just sitting up made me feel that my life essence was slowly dripping out the bottoms of my feet.

The whole experience of being a patient in a U.S. hospital affirmed my decision to leave the United States and work in Indonesia. The singer Jewel has a song, "Life Uncommon," with the powerful refrain: "No longer lend your strength to that which you wish to be free from." That line often played in my head, when I was training and working in U.S. hospitals before moving back to Borneo. Indeed, I had not wanted to lend my strength to a medical system that did not care for everyone, where health care is a corporate enterprise, and where staff are trained not as the healers they are but as cogs in a machine designed primarily to generate vast profits.

After a few months in the United States, both Cam and I were eager to return to Indonesia. However, I ended up having to be hospitalized again en route. Some rough months in Borneo ensued, followed by another hospitalization in Singapore, and then a move to Bali with the plan to just rest and recover. In Bali, my world shrank to the size of a bed, and paradoxically, I began to finally face the bigger picture. I had been focused for many years on the day-to-day work of the two non-profits I helped found, Health In Harmony in the United States and Alam Sehat Lestari (ASRI) in Indonesia, but I had been afraid to really look at the global reality of environmental collapse. Living on the edge of rainforest destruction, I was all too well aware of what was happening, but I had avoided thinking too deeply about it, because the bigger truths of what this means for our planet are so painful. Here is the truth: we are facing extinction of the entire human species and nearly all the biodiversity on the planet. And even worse, this self-destruction is the highest probability of what will happen.

The process of living in a body whose regulatory mechanisms had been screwed up helped me see what was happening to the planet. A healthy body can handle being a little dehydrated or over-hydrated, a little too warm or a little too cold. But when the self-adjusting mechanisms get out of whack, when the parts of the body that balance everything aren't working—well, that's a serious problem. This is what we have done to the earth. There just isn't much of a buffer anymore against the impacts of massive population growth, overuse of resources, plastic pollution, and constant combustion of fossil fuels. We are already beginning to see the results: the sixth great extinction, the acidification of the oceans, the drying-up of water sources, and tragically widening divisions between rich and poor. And just as we take for granted the miraculous systems of our own bodies (until they go awry), we also don't realize how precious and precarious the system is that keeps the earth in balance. Greed has led to the exploitation of the earth for the benefit of very few, and the cumulative damage to the natural buffering of our planet puts us all at extreme risk. But it isn't just greed: my work in Borneo showed me that, for most of us, it is simply hard or impossible to make the right choices. Even when we don't want to, we are all partaking in destructive patterns.

Watching the forests of Borneo disappear sometimes felt like my heart was being ripped out. Over more than two decades, I had seen the fastest rate of deforestation the world has ever known. Even the picturesque mangrove cove where I was stung was bulldozed just a few months later as part of a "development" plan that would probably never benefit anyone. Mangroves are precious fish nurseries that also protect coasts from erosion, and, like all forests, they produce oxygen and regulate the temperature of the earth; but these impacts are never calculated as part of the project costs. Equally enraging was the thought of the real benefits that such an amount of money could have provided in a region where the average income was less than two dollars a day.

For myself and for the earth, I was going through the stages of denial, grief, depression, and anger. Why were we destroying ourselves and why couldn't we each, individually and collectively, make the choices that would improve both our short-term and long-term well-being— and the planet's? Yet like the grace I found in the deepest throes of death, when I truly looked at the pain of the earth and finally let it sink in, I began unexpectedly to have hope for our future. The human species may be deeply flawed, but we also have the capacity for beauty, transcendence, and unexpectedly rapid change. As an exchange student in West Germany, I had gone on a trip to East Germany three months before the Berlin Wall came down. Everyone we met there expressed total despair that anything would ever change. Yet only months later, I found myself with tens of thousands of others whacking away at what had seemed a physical manifestation of political certainty. We smashed holes in the wall until we could reach through and assist East Germans for their first experience of the West. Could the human experience with climate change be the same? Would an apparently unchangeable reality suddenly begin to crumble under massive people pressure?

My fragile hope was buoyed when the results of ASRI's five-year impact survey came in and showed that it was actually possible for humans and the natural ecosystem to simultaneously thrive. We did not have to see these two things in conflict with each other. In fact, if both don't thrive, neither can in the long run. Remarkably, illegal logging rates in the national park where we worked had plummeted, *and* health had dramatically improved. In addition, people who switched to sustainable livelihoods were doing just as well or better economically.

Around that same time, I had a dream that felt prophetic, telling me that I had to choose to work on a much broader scale than just a single location in Borneo—and that my choice, and everyone's choice, deeply mattered. The dream shook me profoundly, as I had no desire to change what I was doing and hoped to go back to help in the clinic as soon as I recovered. But the message was clear: if I did not make this choice, we might not survive as a human species. The dream also made clear that I am not alone in this, that each of us has something to give to the solution. Every person who faces this truth has to ask themselves the same question: What is my personal role in the global transformation toward a thriving future?

The science is clear. We no longer have the time to push this question off to some more convenient point in the future. We are living in the most critical period in the entire history of humankind, and if we do not act now, it will simply be too late. Did I have the courage to accept a bigger calling? Do you?

2

Rites of Passage

The early-morning light shone onto the perfect stillness of the river as it meandered through the tidal mangroves. The light reflected so exactly in the water that the nipa palm fronds seemed to arch down toward their reflections, forming a bridge between perfectly symmetrical worlds. We seemed to float in some halfway space, as though the wooden boat were a levitating craft, with our own doubles sharing this liminal state. I was mesmerized by the effect—and delighted to finally be headed into the rainforest to begin a journey that I hoped would lead to a career studying primates.

Jennifer Burnaford was my comrade in adventure for the year, and the two of us were wedged tightly together on the bottom of the boat with our arms resting on the gunwales. When I peeked over the edge, I met my mirror-self looking up from below—but I had to do this very carefully, because the boat barely floated above the waterline. Jono drove the boat, while Dar perched on the bow, keeping watch for submerged logs or crocodiles. Jennifer and I, both twenty-one-year-old undergraduate biology majors (she from Dartmouth, I from Reed College), would run the rainforest research station for a year and do a joint study on the dispersal of seeds by orangutans. Dar and Jono were two local research assistants who had come down from the research station to fetch us. Over the years, they would become my close friends.

Before coming, I had been surprised to discover that Indonesia is one of the larger countries in the world, comprising seventeen thousand tropical islands, with the fourth-largest population. In an atlas, the Mercator projection makes it seem smaller, but on a globe, which isn't distorted, when I placed my hand on the far eastern side of Indonesia, Papua, my fingers barely reached Sumatra on the west. That same hand stretched from the East Coast of the United States all the way to Alaska. I reflected that we were currently just about where my ring touched the island of Borneo. Borneo is carved up into three nationalities: the tiny country in the northwest is Brunei (still ruled by a sultan), the top third belongs to Malaysia, and the rest, called Kalimantan, is Indonesian.

We had had months of visa delays in the United States, followed by weeks of chasing paperwork on tiny, motorized rickshaws in Jakarta. Finally, we had flown to Kalimantan, arriving in the timber-and-gold-rush town of Pontianak. Streets with wooden boardwalks were lined with shops, some selling pure gold, newly dug up and hammered into bright-yellow jewelry, others displaying pungent dried fish in big tubs or plastic containers of every shape and size. These markets provided evidence for my father's pet theory that the plastic bucket may be the most useful item ever invented.

The two huge rivers that converge in Pontianak were always filled with rafts of cut logs, many of them more than six feet in diameter, being floated down the river from the interior forests of Kalimantan. Everywhere in that delta city, you could hear the *chug-chug* of diesel engines powering the giant, brightly painted houseboats and cargo boats that plied the rivers.

The first leg of our journey south from Pontianak to a town called Rasau Jaya was on a road still under construction. Our jeep bounced over a wavy roadbed that seemed to float on the squishy ground of a drained peat swamp. At the side of the road, men swung mallets to break up rock, and dusty women carried impossibly huge piles of the rocks in baskets on their heads, teetering along until they could dump them out at the edge of the slowly lengthening ribbon of road.

In Rasau, we made our way through the crowd and climbed onto a peeling red-and-blue boat called the *Express*. This misnamed boat twisted and chugged its way through the mangrove swamps, over-loaded with people, baggage, and animals, taking more than eighteen hours to travel about eighty miles as the crow flies, before arriving in the last outpost of civilization: a town called Teluk Melano. Our accommodations were in the *sederhana*, meaning "plain" or "simple." It lived up to its name. The walls consisted of sheets of unpainted plywood, with strategic holes to allow people to peer in; the roof was a scorching sheet of tin; lighting came from a single bare bulb; and roosters crowed right outside the "window"—a square hole in the wall, with a flap made of hinged plywood that could be pulled shut to eliminate any hint of airflow. The beds were full of bedbugs, and there were no sheets for the bare mattresses. Bathing facilities consisted of a bucket of water at the end of the hall, equipped with a plastic scoop. The bather stood in front of everyone, wearing a sarong for decency. The toilet was a hole in the floor over the river, and the food was on par with the other amenities.

The first evening there, we met Jono and Dar. They cheerfully showed us the town, which I estimated to be about three times the size of the northern New Mexico village of one thousand people where I grew up. As we walked through the village, the houses made of weath-ered planks cut with a chain saw, a gaggle of barefoot children running after us, we found people extremely welcoming, despite their initial alarm at meeting outsiders who seemed as alien to them as Martians would likely seem to Americans. When we were warmly invited into people's homes, most had no furniture at all or just a single broken plastic chair. The beds were just thin floor mats woven from pandan leaves; similar mats were used for meals, with everyone sitting in a circle to share the rice, salted fish, and maybe a few vegetables, served on glass plates in the middle of the mat. Bathing was done in the river, and the "necessities" were also done there—or else by just lifting a floorboard, as the homes were stilted: chickens and dogs would take care of anything that landed below.

To take in the view, we climbed up onto a fancy, out-of-place steel bridge that had been built with Japanese development aid. Calling what emerged on either side of this bridge a "road" would have been generous at best—semi-connected mud pits was more like it. Through gestures, accompanied by our few words of Indonesian and their few words of English, Jono and Dar told us about the bridge's construction. Apparently, someone's head was buried under the bridge (the universal gesture of a flat hand drawn quickly across the neck was pretty easy to understand). The Indigenous Dayak people of this region had at times taken heads as part of a ritual belief that the spiritual power of a human could be used to protect an inanimate object. We watched in fascination as hundreds of thousands of giant fruit bats emerged at dusk from the wide expanse of forest and headed out toward the mangroves. In the distance, we could see the majestic rise of the mountains of Gunung Palung National Park, with clouds towering above them. This was our goal, and we stared at it, excited and terrified.

I had not been sad to leave Teluk Melano early the next morning. Each stage of our journey had moved us to a progressively less modern form of transport: big international planes, small twin-engine jet, old jeep, the humorously named *Express*—and now we had crammed ourselves into this craft that was little more than a wooden canoe with an outboard motor. After about an hour on the larger river with those amazing reflections, Jennifer and I sat up straighter as our little sampan slowed to take a shortcut. Jono deftly guided it into a tiny waterway through the mangroves, a detour that would bypass one of the big loops of the river. This place was unlike any place I had ever been. There were the stilt-like roots of the mangroves, an intense whine of cicadas,

plops from frogs and mudskippers disturbed by our passage, and gajil-
lions of mosquitos coming at us through the hot, humid air.

When our boat emerged from the mangrove maze, back into the
main river, we saw jewellike kingfishers zip past and watched our first
proboscis monkeys leaping between trees. Farther along, I spotted a
giant bareheaded black bird that my still-crisp *Birds of Borneo* guide-
book told me was a lesser adjutant stork. A few times, we met a lone
logger poling down the river on a raft of logs, but soon only the forest
remained. As the river got smaller, we passed through a field of spiky
rattan vines. This vine, which is used to make wicker furniture, has
long, dangling spikes with fishhook barbs that will rip through almost
anything. Dar, perching fearlessly on the bow of the boat, used his pa-
rang (Indonesian machete) to slash the tendrils before they got us. He
was only partially successful; but a rite of passage often involves blood,
and here we started to pay our dues.

Then we began to hit the sandbanks. The first one was kind of exciting as we scrambled out and helped push the boat over it. By noon, getting back in the boat was starting to seem like a waste of energy. Soon
the forest completely closed over the river, and we were full-time pushing the boat. Jono took off the motor and placed it where Jennifer and I
had been sitting. The boat was so heavy that I began to scrutinize everything inside it, searching for something to jettison. My own admittedly
heavy gear included everything recommended by those who had done
this job before: four pairs of army-issue jungle boots (because they rot
through in a few months); a strong backpack; twelve pairs of socks; a
good pair of binoculars; silica gel to keep the fungus out of anything
mechanical; and an "everyday" dress for the city that was modest and
cool. (I had searched for months for this dress before finding just the
right thing—only to laugh when I saw the label inside: "Made in Indonesia.") The rest of my bag held a small selection of books. In addition
to Jennifer's similarly heavy gear, there was a large bag filled with five
thousand metal tags to mark trees we had been instructed to bring over

from the United States. Topping it all off were boxes of canned food that we had been asked, via shortwave radio, to bring from Pontianak, and vegetables we had bought in Teluk Melano. (Dar and Jono told us that nothing, except the mail, would be more welcome at camp.) Sadly, there appeared to be little we could dump.

After about four hours of pushing the boat, we reached a tiny sapling platform covered by a blue tarp, just on the edge of Gunung Palung National Park. Two loggers invited us to share their meager rations and made us cups of coffee. I sat slumped in exhaustion on a sandbank, drinking coffee made from about equal parts pounded coffee beans, pounded rice, coconut, and sugar, all stirred together in hot water. It was the elixir of the gods. I stared at my bleeding and torn feet, the Teva sandals now only half strapped on, and reflected that this was a far cry from the high heels and red velvet evening gown I had been wearing when I was offered this position.

THIS WHOLE TRIP CAME ABOUT because I was jealous of a friend of mine who kept getting awards and scholarships to go to China. I had decided that her good fortune was partly due to being so *focused* and that my problem was that I was interested in too many things (a fault that persists). So I decided to choose just one of the many things I liked and to pursue it completely. After briefly considering being an astronaut, I chose one of the literally wildest things I was passionate about: orangutans—the "red ape," or in Indonesian, "person of the forest" (*orang hutan*), who live only on the islands of Borneo and Sumatra.

I had been fascinated by them from childhood, possibly because my mother also had bright orangey-red hair. I once read that the difference between caged chimps and orangutans was that both species could figure out how to pick the lock to get out of their cages but that orangutans could let themselves out and then lock themselves back *in* while hiding the "key" in their mouths, so the keepers would not know they had gotten out. From that moment on, I had been hooked. I imagined a romantic life studying them in wild rainforests.

After making my decision, I wrote letters to the authors of every article and book I could find on orangutans and told everyone I knew how much I wanted to go and study them. Four months later, the latter tactic bore fruit, when a friend of a friend told me that a professor at Harvard, Mark Leighton, had a forest research station in West Kalimantan, Indonesia—and that he hired undergraduates each year to manage the research station.

"Call him!" my friend told me. But did I? No. My courage failed me at the idea of cold-calling a Harvard professor. A week later, the same friend announced, "I talked to Mark, and he is looking for someone right now. I told him about you. *Call him!*"

This time, I got on the phone. After a few phone interviews, Mark asked me to come to Boston for an in-person interview. Twice he had me book and then cancel flights to come out from Portland (this was an era where booking happened well in advance of paying). After booking for a third time, I called him to confirm.

"Can you handle outhouses?" he asked. I laughed, and I told him that our home in New Mexico had an outhouse until I was ten. I wasn't concerned at all about that. He asked me if I was sure I wanted to do this.

"The problem," I responded, "is not that I don't want to do this enough, it is that I want to do it *too much*."

"All right," Mark answered. "Tomorrow is the deadline to cancel your flight before you have to pay for it. Call me tonight. I need to think about this. I don't want you to fly out here and waste your money if it isn't necessary. You said you'd be at a classical concert tonight? Can you call me during intermission?" So that same evening, I stood on a red carpet in the foyer of the concert hall wearing heels and a long red velvet dress (thanks to a free ticket from the Reed Culture office). I inserted coin after coin into the pay phone. Mark picked up the phone and, with no preamble, told me, "Kinari, I've decided there is no need for you to fly out." My stomach dropped as he paused dramatically. "I've decided to accept you without meeting you."

JENNIFER AND I SAT SIDE by side, mute in our weariness, as we helped each other tape our sandals onto our bloody feet with duct tape. Velcro, apparently, was not designed to handle water, sand, and hidden sticks and logs. Of all the things to throw overboard, it was already clear that the duct tape would not be one of them. However, the five thousand tree tags, which would be used to label trees when keeping track of fruiting patterns, were starting to seem less and less necessary.

"Can you believe how generous these people are to make us coffee? It must take them weeks to pole down from here on a raft of logs, and we just shortened their coffee ration by four days. Maybe we should pay them in tree tags?" We both smiled weakly at my joke.

Jennifer called out to Dar and Jono to ask how much farther. "*Sudah setengah jauh,*" Dar answered, with a sideways glance at Jono.

"Did he really say what I think he said?" Jennifer asked me, her voice edged with terror. "Can they possibly mean that we are only *halfway there*? It's already four o'clock in the afternoon!"

Jennifer and I were in our early twenties. We were probably in the best physical shape of our lives (and had perfectly functioning autonomic nervous systems), but we were both close to the end of our ropes.

"I'm afraid so." I sighed. "I don't know if I'm going to make it. I feel like I've run a marathon already."

"No, this is *much* harder than a marathon. In a marathon, all you do is run—not push a boat, swim every few minutes, and walk against the current with hidden sticks jabbing into your feet and God knows what else in the water!" Her voice was starting to get high-pitched. "What are we going to do when it gets dark? What about the snakes? And the *crocodiles!*"

I wanted to calm both our rising panic—but my effort was cut short when I suddenly had to run and find a private spot in the woods. Evidently, the food at the "plain stay" in Melano was catching up with me.

As we pushed on, the forest closed more and more tightly over the river as if we were traveling through a living tunnel, with vine fingers and air roots that would reach down and brush you unexpectedly. By five o'clock, I was more tired than I had ever been in my life. And then I began to have to push at the back of the boat—for obvious reasons. Every half hour or so, I'd pull down my shorts for a quick pause before catching up again to the boat.

The thing about the forest is that it is alive, and I mean *alive*. Like nothing I had ever seen. Every possible space is filled with vegetation: trees, vines, bushes, lichens, mosses, and epiphytes. And every bit of vegetation is covered in ants or insects; clumps of leaves sometimes release clouds of bats when bumped; every dangling vine looks like a snake, and some actually are. The buzzing of the cicadas, croaks of tree frogs, calls of birds, and screeches of the occasional mammal are so loud that you can barely hear someone next to you. At dusk, the forest is so loud it is nearly deafening. And I had no idea what was actually out there. What was just around the corner or behind that log? Spiders ran across my face. It was terrifying enough when I could see—but what about when it got dark?

Night fell unnervingly fast: ten minutes of twilight and then darkness. Barely time to get ready, find a flashlight, and prepare our psyches. Through the forest, an incredibly eerie sound began resounding: *waaaaaah waaah waah wah wah wah.* I assumed it must be a cicada, but it sounded more like a creature mourning a mate, not calling for one. When darkness fell—or rather *rose* from the forest floor into the canopy—I realized I had never been more afraid.

The dark in the forest is so much darker than anywhere I had been before. Even at midday, there is so little light reaching through the multiple layers of vegetation that cameras require a flash. At night, it is pitch, pitch-black without even any visible stars. And out of the black come things reaching and swiping across your face, unknown beings slithering past your legs, and noises—oh, the noises, and not knowing what any of them were. And bites, stings, buzzing insects, spiders scurrying across my body. On and on the four of us pushed that boat. By eight o'clock, though, we simply could no longer help, and Jono and Dar were managing the boat on their own. Jennifer and I were struggling to just keep up and hold the flashlight for them. Together, they were forcing the heavy boat through the shallow streambed, while simultaneously watching out for us. I was awed by their strength and stamina.

The pitch black, the winding river, the exhaustion beyond exhaustion, the fear—it seemed like it would never end. But at 10:00 p.m., sixteen long hours after setting off on the last leg of the journey, we saw a flicker of light in the immensity of darkness. At first, I thought it must be a firefly. But no, it was actually the faintest of lights: one little burning wick of flame on a bottle of kerosene. Then calls of greeting into the surrounding darkness—it was camp! I had never been so happy to see any place on earth.

As we struggled up into the bunkhouse, there was a blur of people and names as we met all ten of the other field assistants. Pak Alex introduced himself, in English, as the camp manager. He smiled warmly at us and said, "You're almost there, just one more kilometer to the camp where you'll be living." Seeing our expressions, Alex coaxed, "I think Cam made spaghetti for you." I doubt that anything else could have persuaded us to stand up again.

Within minutes of starting out from the bunkhouse, we had our first experience with fire ants, whose bite hurts like a wasp sting but does not last as long. They swarm like army ants, and the only way to escape is to run, stamping your feet—but of course we didn't know that, and we got many stings before Alex managed to drag us out of the swarm. Farther up the trail, there was another first: Jennifer was bitten by a leech. In Borneo, there are no river leeches, but there are *land* leeches. And Jennifer got one—let's just say, close to the area of the body where one would least want to find one's first leech.

The path wound along the river, and we heard the snuffling and grunting of some apparently very large animals just beyond the reach of our weak flashlights. Later, we found out they were giant bearded pigs, using their sharp tusks to turn up the earth, but we were too scared to even ask Alex what we were hearing. At several places along the path, we had to cross a bridge consisting of a single plank of iron-wood, sometimes with a rope or stick as a handhold. Jennifer and I inched across the gaps, terrified of falling into the little tributaries at the bottom of the deep, dark channels, spanned by the webs of spiders so large they were said to feed on bats. I swore to myself: *You will not give up. You will force yourself to stay at least* one week, *then you can run away—but not before then!* Of course, leaving would be no easy task either. It felt as if we really had reached the end of the earth, a place so remote and inhuman that it might as well have been marked on a map, over a big blank place, *"Here dwell dragons."*

As we continued walking upriver, out of the swamp forest and onto the alluvial plain, the trees became bigger and bigger. Our flashlights revealed buttresses spanning ten feet or more and disappearing high up into the darkness. We were passing into a universe of giants and foreign beings.

And then, joy of joys, we arrived at something human-made. This building of the Cabang Panti research station is called the "beach house," because it sits on a slight sandbank along a bend of the Air Putih River. On the other side of the sparkling clear water was a big gap where a tree had fallen, opening up something rare and precious—a view. In the morning, that gap would reveal a vertical cut through the forest, but that night, we were pleased to see a spectacular array of stars unfamiliar to our northern hemisphere eyes. Jennifer and I sank to the beach house floor, while Alex ran to the main research building to get the only other member of camp. A few minutes later, Cam appeared with a huge smile on his face. He was tall and broad-shouldered, tow-headed, strongly browed, clean-shaven under a handsome nose, and clothed in a ragged T-shirt and shorts that showed his muscular legs.

All that I knew about him at that point was that he was working on his Ph.D. in tropical ecology at Dartmouth and that he had first come to Cabang Panti after graduating from Oxford to do the same job that Jennifer and I had taken on. Jennifer already knew him because he had been the teaching assistant for a field course she took in Jamaica. For the previous three months, he and Alex had resided alone at camp.

"Welcome to Cabang Panti!" Cam greeted us in his flowing English accent. He then informed us, with a twinkle in his eye, that the spaghetti was only about fifty meters away.

"No way. We aren't moving another inch!" we protested in unison. "Bring it here!"

"But before that, Cam, will you show me where the outhouse is?" I implored. Apparently, my diarrhea was coming back with a vengeance. I followed him into the forest, but after ten minutes of wandering futilely around in circles, with Cam saying, "I know it's here somewhere," I could wait no longer.

Apologizing, I simply squatted, right next to him.

"Nice to meet you, too," I said.

3

The Land of Dragons

After a restless night clutching my flashlight and constantly waking to unidentified squawks and rustles that sounded ominously close, I finally got a few hours of good sleep after the sun rose. I wanted intensely to run away, but leaving felt physically impossible, as my muscles protested even the slightest movement.

Body aching, I managed to get up off the thin mattress on the floor and flip it over a clothesline stretched across the space. Cam had warned us before leaving the night before to hang up our mattresses because otherwise scorpions and other creatures would create a home under them. Charming and disconcerting that this was one of the first lessons he felt we needed to learn.

Jennifer and I dug out some soap from the little bag of essentials we had put together the night before and made our way down to the river from the wooden platform with a tin roof that was referred to as a "house," although it only had two partial walls of woven leaves. We had each bought Indonesian batik sarongs in Jakarta, and taking the advice of earlier researchers, I had sewn the ends together so that it could be worn like a wraparound sleeveless dress for bathing. We tentatively slipped across the stones on the edge of the river with our tender, bruised feet and slid down into the shallow river with a sigh. Lying back and resting my head in the water, I looked up at the impossibly tall trees and could see a group of macaque monkeys making their way from their sleeping tree above the beach house out into the forest. I knew some of the trees were twenty-two stories high, but most of the monkeys were traversing aerial pathways only about ten stories up. I found myself actually having a moment of excitement about getting to see orangutans, but I couldn't imagine overcoming the intense fear of being alone in the forest, even if this might be one of the most beautiful views I had ever seen. I repeated the deal to myself: *One week.*

Moments later, nibbles on my feet sent me bolt upright. Bright orange-and-black-striped fish retreated slightly but quickly dashed forward again for a bite of loose flesh. The glints of their skin competed nicely with the gold accents in my waving sarong. Their bites didn't exactly hurt, but they were enough to get me to finish bathing.

We dressed in our still-damp clothes from the night before, as we had left the rest of our gear in our bags. We had been told to just follow the trail upriver, and sure enough, the main camp was embarrassingly close. Despite the proximity, I didn't regret making Cam bring the pasta to us—although the incident in the forest was definitely something I'd rather erase, and I found my cheeks reddening as we approached camp.

Cam, though, kindly didn't mention it, instead greeting us warmly and offering instant noodles with a little of the vegetables that had been brought up that morning from the field assistant's camp. Our huge bags were piled among the rest of the boxes by the back steps, but Cam assured me that Jono and Dar were taking the day off to rest—other staff had carried them up. Cam explained that the field assistants were already off in the forest doing their "pheno" work, recording the trees fruiting or flowering in about fifty forest plots across the many different ecosystems this research area was famous for: lowland rainforest, peat swamps, swamp forest, montane forest, and cloud forest at the tops of the ridges.

After breakfast, Cam wanted to teach us about the rain gauge and temperature measurements. Each day—*if I stayed*—we would have to record the rainfall and temperature range, as this data would be compared with the fruiting and flowering data. As Cam showed us how to measure the rain from the day before and then flicked out the water from yesterday's downpour, he suddenly froze and looked off into the forest behind the camp. I couldn't imagine what sound he had pulled out of the cacophony of noise, let alone what he saw.

"You are in luck. Want to meet your first orangutan?" A smile spread across his face.

Jennifer and I dashed after Cam as he strode long-legged into the forest. And then there she was! A young female swung one-handed from a tree and peeked over her furry orange shoulder to watch our approach. And then it hit me like a big branch falling from above: I had actually done it. I was here in the rainforest of Borneo looking up at an orangutan who was looking down at me wondering how I had possibly gotten here. I agreed with her assessment. What on earth was I doing in this place?

The young female, deciding that we weren't that interesting after all, climbed high up in her tree and disappeared into the foliage. I felt breathless with astonishment and delight, which Cam thought hugely amusing given how little we had seen. For me, though, the desperation to be gone in a week eased just a bit.

Two days later, Jennifer and I, dressed in our crisp army-issue green pants, still-white socks, jungle boots, and green T-shirts, ventured out for our first days of exploring the forest alone. Cam had trained us in reading the hand-drawn map, and we had each practiced measuring the length of our standard paces. Since orangutans didn't follow trails, we would have to learn quickly how to count our steps and keep a constant estimate of the distance we had traveled in a given compass direction. Those silver tree tags that we had brought in with so much effort should also always be kept in our pockets. If we encountered any flowering or fruiting trees, we would collect a sample, hammer a metal tag into the tree, and then record it in our yellow "Rite in the Rain" notebooks.

We were instructed not to venture far that first day. I teetered across the huge tree that was currently forming a rather steep bridge over the river and began to explore down a few trails. Soon I encountered the largest tree I had ever seen in my life—outcompeting, with its flared buttresses, even the California sequoias I had often seen as a child when visiting my paternal grandparents. Then, not knowing how lucky I was, I saw a huge male orangutan, munching away on the leaves of a giant spiky plant. It reminded me of eating an artichoke, although this plant was clearly not quite as soft, as he kept spitting out balls of twisted fiber. I was so excited that I collected a ball to bring back to camp and stuck it in a plastic bag with two fruits I'd already found. Each evening, Cam told us, he would host a "fruit fifteen," where he would teach Jennifer, Alex, and me about the incredible botanical diversity of these forests. One of the fruits I found was like a green misshapen golf ball oozing black sap, and another had a thick orange rind in a star shape from which hung bright purple droplets from little strings. I imagined that a bird could fly by and pluck these purple treats while still on the wing.

Feeling pleased with myself, I nearly skipped back to camp to show off my finds. Cam's thick eyebrows rose when he saw my ball of fiber, and he gently suggested that, given how closely related orangutans are to humans, it might not be a good idea to get that close to their bodily fluids—disease transmission could easily ensue. "And better get rid of that green fruit, too. It's the fruit of a rengas tree—related to poison ivy and even itchier."

Chastened, I shook them both out of the bag over the low leaf wall and made my way down to the dock. The crystal-clear river was deeper here. As I knelt to wash my hands and fill my water bottle, hundreds of glittering fish of at least five species came up, wondering if I might have food to offer them. Later, when we did the dishes, I probably would.

Soon Jennifer was back as well, and her orangutan encounter had us laughing so hard at the table that Cam kept getting distracted from his botanical work of identifying dried specimens. Jennifer recounted how she found a large alpha male who was approached by a young female who seemed to want to mate with him. He appeared totally uninterested and kept looking away—even when she began giving sustained oral attention to his penis. Eventually he gave in, though, and had sex with her.

Although I was still terrified to be in a world so inhuman, the universe seemed to be conspiring to pique our interest, curiosity, and delight on that first day. And a month later, when two of the staff prepared to do the routine boat trip down to Teluk Melano, I found myself only briefly considering joining them.

OUR CAMP WAS NESTLED WITHIN the exquisite lowland forest at the base of the mountain. Farther to the west, there were peat swamps on sandy soil and then swamp forest where the water could be even chest height. Up the mountain, the trails ascended two steep ridges, with a deep valley between. Going up the ridges, the trees get progressively shorter. Cam had recommended we check all the plots that the field assistants monitored before beginning following orangutans so that we could learn all the trails and habitats. On my first foray to check the plot at the top of the mountain at around one thousand meters, I was amazed to discover the moss- and orchid-draped cloud forest was so stunted, I could peek over the canopy just by standing on a rock next to the pig migration trail I had followed.

Any big change begins with apprehension, and three weeks in, I seemed to have slowly edged my way through it, as I started to figure out how to keep myself physically safe and learn the lay of the land. However, spending all day, every day, alone in the forest was showing me that there was something even more dangerous out there than the wilderness: my own brain. Before leaving the United States, I talked to researchers who had done this job previously, and Cheryl Knott told me that I would go through every memory I ever had. I was incredulous, but her prediction had already come true. No matter how far you travel—even to the land of dragons—you can't escape your own soul. *Especially* here, where there were no other distractions.

I had grown up with hippie parents in a progressive community in northern New Mexico, where social mores were being torn down on principle. My parents both had Ph.D.s in psychology and had met in graduate school in New Orleans, where my mother grew up. They traveled around the world, conceiving me in Afghanistan and picking my name up in India. I was born in Michigan during an ill-fated attempt by them both to be professors, before they dropped out and followed a friend to New Mexico. For a while, they both became jewelers and farmed garlic, but my father soon got a job working at a community college. Eventually, he moved on from that position and became a data scientist at Los Alamos National Lab—famous for being the place where the atom bomb was developed. My mother spent her life creating beauty in various forms: jewelry, stained glass, painting, gourmet food, writing, and architecture (she built six houses over the years).

By the time I was seven, my parents were divorced, but my father bought the property next door, and we would often duck through the plum hedge to visit him. Dad could expound on just about everything (incredibly useful in the era before the internet): evolution, human history, mathematics, astronomy, identification of fossils, and how to build or fix things. He was also acutely conscious of raising girls in a patriarchal world, teaching my younger sister and me to change our own oil and tires, and instituting rules like, "If you can't open it, you can't have it." We learned all kinds of tricks for getting lids off jars.

I was free to run and roam, to ride a horse or hike up into the hills in the morning and come home at night. Dixon, New Mexico, the town where I grew up, lies on the ribbon of life called the Embudo, a wild, clear trout river that cascades down from the snowmelt of the Sangre de Cristo Mountains. The river feeds the several-hundred-year-old acequias that irrigate the valley, creating a dramatic distinction between the arid high mountain desert landscape and green lush river bottoms.

My happiest memories were of clambering over granite boulders in our favorite deep canyon, with the rustle of the wind flowing through the yellow leaves of cottonwoods; the tart sweetness of wild grapes; the rich smell of sage; the almost drinkable smell of living water in a dry world; and the splash and laughter of friends as we played in the waterfall or lay naked in the hot sun. In many ways, I felt raised by the land herself.

I grew up knowing, unquestioningly, that there were many paths in life. I learned that any activity—building houses, making art, or shaping the course of a life—can be done in many unexpected ways and that most people can learn to do almost anything. For example, when I was eleven, after a fight with my sister, I stormed in to my mother at her jewelry bench and said, "I just can't do it anymore! I need my own room!" To my great surprise, she sighed and told me she knew this day was coming—but that I would have to help build it. And I did. By the next week, I was digging the foundation, and I assisted with every part but the roof.

But not all the memories rising up in the forest were positive. Dixon had its advantageous sides, but it was also an extremely dangerous place for children to grow up—and not, primarily, from the horses. Drugs, alcohol, wild parties, and extremely minimal oversight were parts of life. I had been surprised in college to find few of my dormmates had friends who had died or were imprisoned. I had many. One of my sister's two best friends was murdered, and the other died in a drug-associated car accident after getting out of jail. We did our best, but I took on many of the norms of my community, including promiscuity, experimentation, and high-risk behavior. For reasons that were somewhat unclear to me, of the many men I dated, I was nearly always the one to leave—and often not kindly. Nor was I always respectful to those around me. I even betrayed my best friend and slept with her boyfriend.

In the silence, I was forced to face these shadow parts of myself. Many of these memories—and others that peeked around the edges but which I feared to look at too closely—brought up a toxic mix of shame, fear, and anger. In some cases, the shame was legitimately mine; in others, it was the sticky slime from others' behavior that wouldn't wash off. Sometimes I had made poor choices; other times, I blamed myself for things that were not my fault. And I couldn't always distinguish between the two.

In addition, life at home was often extremely difficult, with storms of emotions raging. My role in the family was to provide the emotional ballast and protect my sister (not terribly successfully). The few times I tried to escape this role, my father forced me back and told me it was my job to take care of my mother. Finances were also usually quite strained, and Mom would yell at my sister and me for drinking too much milk because she couldn't afford it. I once argued with my father for over an hour, as he would not pay for the windshield wipers that were needed for the car I used to drive carpool. It was not until I went to Germany as an exchange student that I got to experience for the first time what it was like to just be a child and be the one cared for—emotionally, physically, and financially. This experience helped me see that there were actually other choices in life, and by the time I got to college, I was done with drugs and had one stable boyfriend for three years—although I broke up with him and immediately slept with another man shortly before I left for Indonesia. Another moment to not be so proud of.

Weeks of simply sitting in the pain of these memories followed. I found it remarkable that I had managed to avoid really thinking about any of these things for most of my life, focused on getting through the next day and finding ways to distract myself. Yet here in the forest, memories kept crowding in, unbidden. One day, I was thinking about my best college friend, Preetha Rajaraman, who came to Reed from Botswana, and how I would like her to be my first child's godmother. And then, all at once, I was overwhelmed by a relatively recent memory that I had somehow completely blocked out. The day before I left for Indonesia, I had had an abortion. I'd always been adamantly pro-choice and also couldn't imagine that I would ever make that decision myself. Yet when I found out two days before I was supposed to leave that a dalliance had resulted in a pregnancy, I hadn't felt a moment's hesitation. I would *not* give up my adventure in Borneo!

For me, the choice had been pure selfishness. Most people I knew saw this issue as purely black or white, but in the forest, the profound grayness overwhelmed me. For days, I just cried, tears dripping onto my Rite in the Rain notebook. There was nowhere to run. No activities to distract myself. I simply had to be in the pain.

While during the day, I was besieged by these swirling memories and thoughts, at night, I was more literally besieged. There was an army of cockroaches in the camp, and over these, I decided, I had a chance of prevailing. The forest roaches, with their translucent wings and long antennae, were more pleasant than city varieties—but I still didn't want them scurrying over everything. At first, I tackled the storeroom (shelves open to the forest), but after a series of lost battles, I retreated to the wooden planks we used for a kitchen. Within days, it was clear that, too, was hopeless. Next, I staked out the table where we ate dinner by placing bowls of insecticide under each leg. But that night, by the light of our two candle stubs, Cam, Jennifer, Alex, and I could still see the cockroaches slowly approaching through the darkness, hoping to snatch something off one of our plates while we weren't looking. Then I made a discovery: by smashing my fist down onto the table, I could send the cockroaches scurrying back to the dark edges of the table. It was an ignominious end to my "war," and I was pretty sure my dinner companions didn't enjoy the sound of my fist smashing down every few minutes. However, I was determined to hold one sacred core: no cockroaches on my plate *while* I was eating!

Just as my nighttime war failed, eventually my internal defenses also crumbled, and I had to let the cockroaches of my mind feast on my core understanding of self. I had been forced to face myself, and I was not impressed. Tromping through the forest, I began thinking of everyone I considered an exemplary human being—or even just a bit better than average. My high school biology teacher had actively combated racism by seeking out smart Hispanic and Native American students and encouraging them to sign up for the honors classes they might not otherwise take. He even got all of us to come in on weekends to work on our science fair projects. He had compassion and respect for his students, and he made science so interesting that we got hooked. What made him so different from my easily annoyed math teacher, who didn't seem to care whether we learned anything or not and once told me he would give me an A if I kissed him? And how could Gandhi give everything, even his life, to the cause of freeing his country? Or Dr. Martin Luther King Jr.?

These people sacrificed so much and made such a huge difference in the world. What motivated them to make altruistic decisions? My mother believed it was all genetics and that changing behavior was impossible, but that didn't feel right to me. Was it possible to become a better person? What did it even mean to be "better"? Was all behavior just culturally dependent? Was it acceptable for people in Teluk Melano to murder someone to have a head to put under the bridge, if that was part of their traditions? Was my own behavior forgivable if it was normal for the community in which I had been raised? Yet some of the things I had done just didn't feel right. When I reviewed my life, I realized I had been fully willing to hurt others for my own interest.

I began chewing on the idea that, in my own life, the essence seemed to be selfish behavior versus non-selfish behavior: choosing the good of others over one's own pleasure, comfort, ease, and safety. Where did the people I admired find the strength to make those hard choices—those non-selfish, loving choices? And why had I made so few of those choices in my life? My very core felt fundamentally flawed. Was there a way for me to change these things about myself? Could facing these demons allow me to make different choices in the future? I hoped so.

4

Primate Love

Four hours. *Four hours* of sitting and recording in my notebook, every five minutes, that the two wild adult orangutans I was observing were still eating the same *Alangium* fruits. Maybe I didn't want to be Jane Goodall after all.

I had gotten up about an hour before dawn and walked from the research camp to the site where the pair had made their nests the night before. Traversing the dark forest by flashlight, I noticed how much calmer my soul felt than in those first two months. The deep pains began to lose some of their grip. Often, I found myself simply experiencing the forest.

By the time I made it to the trees the orangutans had nested in the night before, they were already up and eating fruit in the nearby *Alangium* tree, although I could only make out black shadows moving against the lightening sky. The night before, the male, who Jennifer had named Xavier, had made his nest close to the trunk of a big tree. Demonstrating incredible strength, he had broken large branches down to make a nest that would support his weight of about three hundred pounds. His mate, whom we called Kristen, at a third his weight, had nested higher in a smaller tree, where she had made a bed big enough for herself, her nursing baby, and her older child, who I thought might be between four and six years old.

The fruiting mast was beginning. Southeast Asian forests, unlike northern forests, only flower and fruit approximately once every four years, probably triggered by El Niño events. This cycle means that orangutans go through three lean years before getting a bonanza of fruit. Initially, the orangutans we followed were mostly solitary, and they ate pretty unappetizing things like bark, termites, and the bases of epiphytic leaves. Now they had a more appealing diet, and I was also trying out many of the same fruits. Curious to know what they tasted like, I scavenged a few *Alangium* fruits that had been knocked loose by the activity above. This annoyed the orangutans; Xavier broke off a branch and threw it at me, while Kristen let out a few threatening "kiss-grunts," smacking her lips and emitting a deep guttural rumble.

"Sorry, sorry!" I muttered, with my head down. The orangutans tolerated being followed, but they clearly didn't like it. More than once, I had been given the slip by a clever orangutan. Now I sat down on a log with my stolen fruit and decided to have an early lunch, although it was still only about ten o'clock. I set down my compass and notebook next to me to rummage through my backpack. My lunch was our standard: a few crackers and wedges of canned cheese. Comparing the many pieces of fruit the orangutans were eating with my small selection, it was clear which meal had more calories. Although it's also true that I wasn't trying to put on extra weight to make it through four years until the next mast.

After eating a few crackers, I squeezed the fruit, and out popped a single seed coated in clear pulp that tasted like a tart grape. I was always mindful of Cam's warning to try only the fruit I saw an orangutan eat— others could very well be poisonous. With an estimated five thousand native tree species in Borneo (compared to just thirty-four in England), trusting the orangutans was the best way to know what I (also a primate) could safely consume. The fruit tasted luscious after months of a very limited diet. Our entire food repertoire consisted of rice, salted fish, pasta, tomato paste, canned cheese, crackers, instant noodles, peanuts, and occasional homemade fry bread (with flour weevils adding a little extra protein). Boat trips would mean vegetables, but the last trip had been two long weeks before. For my cooking rotation the night before, I had used the last of the cabbage. The first step was to scare away the cockroaches, then take the black rotting ball down to the river to wash off layer after layer of slime, slowly peeling away the decayed cabbage to uncover an edible core. Fried with the last of the garlic and peanuts, served over rice with sweet soy sauce, the cabbage had been surprisingly palatable. But now we had no vegetables for another two weeks until the monthly trip went down.

As I sat gnawing on the *Alangium* seeds, I glanced down to see the cracker crumbs on the ground between my boots marching away. It still amazed me that within seconds of the crumbs hitting the ground, ants would find them. This says volumes about how precious resources are in the forest. Even a fallen tree can be devoured in weeks by termites and fungus. Latrine pits were always empty—within minutes, dung beetles would arrive to roll up balls of "nutrients" and take them away for their larvae. With the fruiting season coming on, the forest seemed to be rejoicing in the extra food—but not so much that cracker crumbs would go to waste.

New noises attracted my attention. Xavier was approaching Kristen. Her older child was sitting and eating in some higher branches, while the baby was testing out his branch-swinging skills. It was clear that Xavier had an erection, and he sat on a branch slightly lower than Kristen and gently pushed her legs apart, beginning to explore her genitals with his mouth. Since that first follow of Jennifer's, we had both gotten pretty used to observing lots of sex, including oral sex. For the last month, Xavier and Kristen had been traveling together, and they mated at least two or three times a day. At this moment, Kristen was not playing hard to get; her tilted head, rolled-back eyes, and splayed posture were comically human looking—though, as a good behaviorist, I knew better than to anthropomorphize. Soon, Xavier climbed up onto her branch, and after a few minutes of exertion, he leaned back against the trunk with a few grunts. Again, draw your own conclusions. Checking my watch, I made dutiful notes in my notebook.

After sitting calmly for a while, Kristen decided to move to the next tree. She gathered the baby into her arms and made a kind of click sound to get her older child's attention. Given that Kristen and Xavier had been traveling together for months, I wondered whether he was making sure that her next baby would be his—and that made me wonder if he knew that the previous two were his as well. Xavier took a few more minutes leaning against the trunk, looking remarkably satisfied, as I packed up everything and prepared to follow. Males tend to travel lower in the trees, where the branches are stronger; the tricky part about following the two of them was that, to keep an eye on Kristen, I ended up having to stay much closer to Xavier than I would have liked. I took Kristen's compass direction, and as we moved off, I counted my paces. Counting paces and keeping track of the compass directions had become so automatic that I could think about other things at the same time.

Before my thoughts could wander too far, a sound like cascading hail told me I had another event to record. As Xavier moved on, I found the spot below his last perch and squatted down to begin counting well-cleaned seeds. Orangutan poop is almost exclusively seeds—an unintentional service of dispersal that they provide for the rainforest (well, unintentional on the part of the orangutan, but very much intentional on the part of the tree). This process, in fact, was what Jennifer and I were studying. Xavier had just spread more than one hundred *Alangium* seeds over about five square meters. But I didn't know if these seeds came from the tree Xavier had just fed on or from the *Alangium* tree they had visited yesterday; only when we witnessed them feeding on a unique species of tree over a three-day period could we calculate dispersal distances.

So far, we were finding that the orangutans dispersed seeds very far (up to six hundred meters, or almost two thousand feet). Germination trials comparing orangutan-pooped versus macaque-spat seeds suggested that, for certain tree species, only the seeds that had passed through the gut would germinate. This finding indicated that if the orangutans were to go extinct, they might take some tree species with them.

As I quickly finished counting, I had to jog a little to catch up with Xavier and Kristen. Ducking under a twisting liana, I emerged into a rare sunspot of light shooting down through the thirty-plus stories of the forest. It was filled, as usual, with butterflies using the magic of the light to show off their exquisite colors. And there was my favorite butterfly, of the Latin genus *Idea*. These are white, tissue-paper-like butterflies with tiny black spots that flutter lightly, the way ideas flit from one topic to another, with almost no physical presence, dancing up into the high reaches of the forest and then down almost to the forest floor, never stopping. I once sat for an hour and a half, focused on one of these butterflies, certain that it *must* land to rest, but it didn't, constantly fluttering, catching the light and jumping from one flight of fancy to another.

I didn't know where we were going, but, judging from my experience in following the orangutan pair, I knew Kristen did. Sometimes we would travel for forty minutes without passing a single tree of the fruit species that was most ripe that week and then suddenly come upon one. Later, I would realize that Kristen had made a beeline between the only two trees of that species in the area. Orangutans must have an incredibly complicated map in their heads of the exact location of every tree of every edible species—and when it is likely to fruit. Kristen was imparting that information to her children as they traveled, and she had probably learned it from her mother. I just hoped the trees would continue to be there for the next generation.

Farther up the hill, the orangutans skirted around the top side of a gap left by a tree that had recently fallen. Gaps create openings through the dense forest that allow for views into the surrounding trees—and,

on a slope, sometimes the view extends into the canopy or even beyond. This view was spectacular. Kristen and Xavier stopped to rest as
well, perched on the top side of the gap and looking out over the magnificent forest that stretched to the horizon, wisps of mist rising from it.
Someone had told me that an orangutan could travel from coast to coast
on this, the third-largest island in the world, without ever touching the
ground. Looking out over that expanse of forest, I could believe it.

I had recently learned a new phrase in Indonesian: *"Tak kenal, tak
sayang."* If you don't know something, you can't love it. As I was beginning to understand the forest, I was indeed falling in love. During those
terrifying first weeks, I had felt immersed in what seemed like ubiquitous green. But now, looking at the tree crowns across the gap, I didn't
see *green,* I saw a rainbow of colors. Each tree was a different shade—
aquamarine, dark emerald, soft moss, bright spring green, silvery shim-
mer, yellow, deep purple green, even reddish copper. And their leaves
were so different. After I learned to really examine what I was looking
at, I realized that I almost never saw two trees of the same species at one
time. And it wasn't just the trees: the insects, the mushrooms, the orchids,
the birds, the butterflies, the frogs—everything was mind-bogglingly
diverse. Collecting, collating, and trying to understand so many types
can be captivating.

Extreme diversity means that *everything* is rare and that you are
constantly seeing things you will probably never see again: a fluffy
white moonrat, the size of a house cat; a sun bear with huge long
claws; a ground squirrel standing as tall as my mid-thigh; a tiny bird so
brightly colored that it could have been painted by Gauguin; a green pit
viper on a branch that I *almost* put my hand on; and a troop of half-inch
iridescent blue ants attacking a foot-long purple millipede. The birds,
of course, could be spectacular, like the paradise flycatcher, whose two
long dangling tails rippled through the gaps, catching the light like
opals. But over time, even the secretive brown birds enticed me with
their melodious calls and lured me in to learn their names.

The crack of a branch told me that Kristen had decided to move on; she always led, and Xavier came behind. After a few minutes of following them, with all of us grunting up the steep hill, my mind stopped being so present and became distracted. Like the *Idea* butterfly, my thoughts kept fluttering. First the small worries. We would need to paint the stilts of the main camp with diesel fuel, because the termites were invading again. The last time they got into the library, we lost twenty of our precious books before we noticed anything. And we still hadn't received the money from Harvard. Our field assistants were not going to like it if we didn't pay their wages for another month, and I didn't want to have to borrow from Cam again. Cam and I had gone out a few weeks prior to Pontianak to replenish stocks and pick up money. But our boss at Harvard, Mark Leighton, had only sent enough to pay off our debts. When we called him, he promised to wire more in a few months, but I didn't know what we were going to do in the meantime.

The trip out to Pontianak had been steamy in more ways than one: being crammed tightly next to Cam in the boat, discovering avocado-and-chocolate smoothies, and a particularly intense moment when I ate my first durian. The rich flesh of this smelly fruit has to be sucked off of the large, smooth seeds, and Cam was clearly getting uncomfortable as he watched me do this. That night in my hotel room with its open ventilation slats onto the hall, I had seen a shadow pacing back and forth in front of my room and wondered if it was him.

Sometimes I felt like the forest was a giant curved mirror, reflecting back one's emotions, magnified and distorted: one moment joy and the next fear. My major concerns fell into three fairly generic categories: What was I going to do with my life? What was the meaning of life? And what was I going to do about Cam?

Furthermore, I wasn't at all sure that I still wanted to be a primatologist. I loved the forest, but I missed interacting with other humans. And sometimes these orangutans could be downright boring—they would sit in a tree for hours on end doing nothing: not eating, not mating, just sitting. Maybe they were thinking profound thoughts, but I would never know. Maybe I just didn't have enough patience.

If not primatology, what should I do with my life? As we hiked higher up the slope, I began to hear my least favorite sound: the whine of a chain saw in the far distance. What good would it do to prove that orangutans played a critical role in sustaining these amazing forests if there were no forest and no orangutans left? I waited for the next sound, knowing it would come: the enormous boom of a giant dipterocarp tree hitting the ground. Sometimes it wasn't exactly a sound but rather a crashing vibration that you felt in your feet—a vibration that made my heart ache for the loss of those exquisite rainforest giants. Would I be able to tolerate going into conservation? It seemed too hopeless and painful to bear.

The previous week, I had had an interesting conversation about illegal logging with the field assistants, when I went down to their camp in the evening to teach English.

"Kinari, it's *impossible* for all this forest to disappear," Mourni assured me. "We could log forever and not cut it all down. If we cut down one part, the animals will just go to another part. They will be fine." Hermanto agreed with him.

Tadyn, who used to work as a chain saw man on a logging crew before his back was badly injured by a falling tree, looked less certain about not being able to cut it all down. But he had another argument. "What will the local people do? There are so few ways to make money here. And if someone gets seriously sick, you know we have to travel for three days, and pay a lot of money, to get to the only decent health care, at Serukam hospital. We are all very lucky we have this job, because there are so few ways to make money around here except logging."

Listening to them opened my eyes to a new way of seeing the logging. Even for subsistence farmers who grew their own food, money was necessary for some things. The main things you had to have cash for were weddings, building houses, and health care. The first two could be planned for, and they could always wait until a good crop of rice came in or a big fruit season. But health care was different. When you had to have it, you had to have it *then*.

When Cam and I were in Pontianak, he had taken me on a side trip to see this hospital that Tadyn was talking about. Cam called it a "miracle in a valley." We flew there on a tiny, single-engine plane. There were four seats, including the one for the clean-cut pilot, and we had to be weighed, as well as our small backpacks. Cam helped me in as I held my green everyday dress down with one hand as it was blowing up from the propeller. After we got in and the pilot started checking all the instruments, I asked him, "Wait, aren't you going to give us a safety briefing?"

He looked at me with wrinkled brows. Then he glanced around the plane, clearly trying to think what to say. With a smirk, he pushed down on the lever that opened the flimsy plastic door: "This is how you open the door." He paused. "Yep, that about covers it. Let's go."

And off we went, flying low for a few hours over vast stretches of forest and the occasional tiny village nestled in a valley. When we finally got close to the hospital, the pilot made a wide circle, and we could see white buildings spread out over a lovely compound. The most prominent feature was the grass landing strip. Even from the air, we could hear the wail of the siren announcing the plane's arrival. Some people walked out of the hospital to meet the plane, while others dashed onto the airstrip to chase the chickens and goats off to the side. I'd never been afraid of flying, but that landing was a bit nerve-racking. As we bounced along the ground, I consoled myself that at least I knew how to open the door.

Moments after coming to a stop, I was introduced to Margie Geary, the Minnesotan nurse who cofounded the hospital with her physician husband, Wendell Geary. She greeted me with the first of many warm smiles I received there. I had never met friendlier or more loving people than I met there at Serukam hospital—both the Indonesians and the few expatriates.

Later that day, while getting a tour of the hospital, Wendell asked if I would be willing to join him in surgery, to act as the bookholder, since it was useful to have someone who could read English. Although a general practitioner by training, Wendell was famous for his surgical skills. Still, he liked to have the reference book at hand when he was doing surgery. I was surprised and nervous to be asked, but happy to be of help. Positioned with the book on a step stool next to the nurse anesthetist, who was squeezing air into the patient's lungs with a bag, I watched Wendell make that first incision for a mastectomy. I was both horrified and fascinated, and, with a few deep breaths, my light-headedness passed. Then it actually became fun to read through the procedure and say things like, "Okay, now it says to watch out for the medial pectoral nerve." Wendell murmured to himself as he poked about, before announcing with satisfaction, "Ah yes, here it is. Good. Now what do we do next?"

As I walked out of surgery, hoping that the procedure would save or prolong this woman's life, the Indonesian doctor who had been as-sisting Dr. Geary asked me if I was planning on becoming a doctor, too. The thought had never crossed my mind, but I surprised myself by smiling at her and saying, "Maybe." It was such a strange thought; I didn't know any doctors personally, and I had never had any inclination in that direction, but now I was thinking that there would be something very satisfying about helping people so directly. Something about read-ing the instructions also made it feel less of an impossible thing to save lives.

Yet looking around the exquisite forest, I wondered, *How could I possibly give this up?* It was clear to me that planetary well-being re-quired both saving rainforest and helping people, and work was needed intensely on both fronts. I was also beginning to think, thanks to the local staff, that these two problems were connected; Tadyn's story about logging to pay for a hospital visit suggested that to save rainforest you might have to provide people with affordable health care. Perhaps the best way to save orangutans would actually be to save *people.*

At Reed, I had learned both the idea construct called Maslow's hier-archy of needs and the Blackfoot First Nation's beliefs upon which it was based (although Maslow didn't acknowledge that[1]). In Maslow's hierarchy, the most basic needs are physiological—air, food, health, shelter—followed by safety, love/belonging, esteem, and finally self-actualization. The Blackfoot see things similarly, but for them, the personal level is just the base—the bottom of the tepee. Beyond that comes community actualization and then long-term community sustainability and connection to the Oneness of All. First, people needed their basic needs met, and I understood that for many people around Gunung Palung, that was not the case. Once that happened, as it had for me, one would have to be brave enough to do the personal healing necessary to face one's shadows. I was still working on that. But this personal work had to happen within the wider community work of healing. The Blackfoot, like my Indonesian colleagues, didn't see people as only individuals but rather as part of a larger whole. Living in the rainforest, that felt painfully clear—my very existence was dependent on the rest of the team and on the forest itself. The part that really bothered me, though, was the Blackfoot pinnacle stage. It was the top of the tepee where the light shone in and the smoke was let out: that idea of Divine Oneness and perpetuity of community—sustainability in its greatest sense. It was the acceptance of all of existence as one—the individual nested within the community and the community nested within all of reality. The problem was that I had been raised a devout atheist. My parents were basically evangelicals believing in proselytizing the truth of no God. And I had always agreed with them, absolutely rejecting any belief in anything bigger than what was clearly visible and measurable.

This question of something greater than myself also tied into my thoughts about trying to become a better person. Amazingly, I had found a book on exactly this topic, tucked away in our small library at camp: Dr. Martin Luther King Jr.'s book *The Strength to Love.* As with

---

1 https://lincolnmichel.wordpress.com/2014/04/19/maslows-hierarchy-connected-to-blackfoot-beliefs/ and https://www.blackfootdigitallibrary.com/digital/collection/bdl/id/1296/rec/1

many of the books at camp, I had to guess at some of the words that had been obliterated by termite tunnels. It bothered me that King placed the source of strength squarely in faith in God. He even argued that without faith and the assistance of the divine, truly loving behavior is not possible. I went through the role models I had been thinking about and was disturbed to discover they were all people of faith. They came from many different religious backgrounds, but they all focused on the sacred.

And then there was that annoying Cam. When we discussed faith and science, he asked me, "Where is the proof that God does not exist?"

I had countered, "And where is the proof God *does* exist?"

"Well, if we had no proof in either direction, how should we make the decision?"

Cam's question was not rhetorical. I could see that it plagued him, too—although there was likely an element of playful teasing as well. But his questions had an uncomfortable knack for sticking in my head and festering. Thinking about Cam led me to the next worry. I liked him. I liked him a lot. And despite our inauspicious first meeting, his interest in me was obvious. But it was extremely clear to both of us that a rainforest romance, in a community of four people, was a very bad idea.

Cam had a British father (who had grown up in Tanganyika) and an American mother, but he had grown up typically English—including being sent to boarding school when he was ten. Cam's grandmother, a Cockney from South London, had spent her life rising above her class. Like Eliza Doolittle in *My Fair Lady,* she changed her accent and married up. She convinced her husband to sell their small farm in England and buy a huge one on the slopes of Kilimanjaro, thereby through colonialism sidestepping the class system of England. From her perspective, she achieved success when her grandson was admitted to Oxford. That grandson, though, had almost no interest in wealth or status, and he kept running off to remote places like Borneo. After spending a year doing the job that Jennifer and I were now doing, Cam had decided to pursue a Ph.D. in tropical ecology at Dartmouth. His passions were plants, philosophy, and wild places.

Curious to know what kind of person he would be spending six months with in such intimate confines, Cam had tried to find out about me before I arrived, writing everyone he could think of who might know someone who knew me. He had confessed this to me on our trip back into the forest and told me that the words that came back were "mostly harmless." I had to chuckle at the thought. Yep, that probably summed me up. Fairly benign most of the time, *but don't trust her for a second.*

Like Xavier with Kristen, Cam seemed to be following me, psychically. And in my search through the forest for new fruits of thought, I kept finding myself looking over my shoulder to make sure his powerful presence was still attentive. But I was deeply conflicted about really turning toward him and what that would mean. Cam had made it quite obvious, in talking about relationships in the abstract, that he was only interested in starting something that had the potential to be serious. And I, for sure, was not looking for marriage. Nor was I sure I was even capable of a respectful long-term relationship. I was also determined to try to be a better a person—which in this case, meant staying away from him.

Pulling myself back to reality, I glanced down at my big Ironman digital watch and realized I should soon hear Jennifer's owllike hoots as she would come looking for me to change places. A medium-pitched *whoo-whoo* call carries well, and this is the method Indonesians use to locate each other in the forest. I hoped she would be able to find me, as we had traveled quite far from the map location I had given her the night before, indicating where the orangutans nested.

The plan was to switch off around noon each day. The one on duty would then follow the orangutans until they made a nest at night, and then go out again before dawn. Depending on how far away they were, that meant rising between 3:00 and 5:00 a.m. She should be finding me soon.

An enormous crash made me freeze, one foot off the ground, instantly stopping my wandering thoughts. Xavier had been using his typical traveling method, leaning forward on the tree he was in and bending it far over, then grabbing another, and using the force of the swing to be carried on to a third tree. But this time, as his tree bent forward to its maximum, it suddenly broke and Xavier smashed thirty feet to the ground, landing behind an enormous fallen log.

With my mind drifting, I had been following way too close to Xavier, and now he was only about ten feet in front of me. Thank God (except there wasn't one), the huge tree trunk lay between us: facing an angry male orangutan on the ground was not something I ever wanted to do. I couldn't see him at all over the log, and I remained perfectly frozen, just bringing my foot *verrry sloowly* down. I waited and waited. The only sound I could hear was Kristen coming back and smacking her lips with concern at Xavier as she leaned out over a tree fern to get a view of him. From behind the log was total silence. Nothing. No grunts, no heavy breathing—nothing. He must be dead. I waited another minute to be certain. Not a sound from Xavier. I decided to check it out.

I lifted a leg to step forward, and in that same instant, I heard a rustling in the trees behind me. My head spun around—and there was *another* male orangutan! In a second, I understood: he must have been following behind us, out of sight, and now he saw his chance with Kristen. A story Cam had told us flashed through my mind, about an observer getting caught between two male orangutans. Those orangutans chased each other until one of them switched and went after the researcher, who ran flat out until he tripped. The male orangutan jumped on his back, growling and snarling, while the man played dead and prayed for his life. I had wondered out loud how anyone could be so stupid to get caught between two male orangutans. Now I understood. But luckily for me, one was dead. Evidently, this second male had come to the same conclusion at exactly the same moment, and we both made a move simultaneously.

But we were both wrong. With a roar, Xavier leaped up from behind the log and seemed to fly into the trees. Our matching expressions of total disbelief would have been comical—except that I was smack between two male orangutans that were clearly about to fight.

I flipped ninety degrees and ran as fast as I could. I leaped over another treefall before spinning around to look back. Xavier was tearing and roaring through the trees after the other male, who was trying so hard to get away that he fell to the ground just a short distance from where I had been standing. Xavier jumped from above and—*boom!*—crashed down onto the second male. They were snarling and ripping and rolling. One of them—I couldn't tell which—picked the other up over his head and pounded him down onto the ground with a cracking thud. The one on the ground tackled his foe. Biting and snarling, they became one mass of rolling anger and flying blood. They tumbled down the hill beyond my sight, but I could still hear the awful booms and crashes as bodies were smacked into trees or onto the ground. From above, Kristen had come back and was following closely, kiss-grunting and throwing branches down at them.

Finally, after a truly sickening thud . . . silence. I waited a *long* time. This time I was not stupid enough to go check if either of them was still alive.

Then, slowly, as slowly as I had ever seen an orangutan move, Xavier climbed up a thick vine to a towering dipterocarp tree, high above the misty rainforest. He sat down heavily on a huge branch that was larger than most trees. Kristen climbed up and sat down next to him, gently placing her hand on his knee.

"Well done, my love," she seemed to be saying. "I knew I chose the right guy."

At least she knew what she wanted.

5

Waking to the Gibbons' Song

GUNUNG PALUNG NATIONAL PARK:

EARLY DECEMBER 1993

I lay on the thin mattress on the floor of my little hut, having just awakened to the soft patter of rain on the tin roof—and that moment of sheer joy when I realized that this was a rain day. About once a month, instead of the usual afternoon drenching, the day would begin with a drizzle, which invariably meant it would rain all day. The tradition of the research station was that these were the only days we ever had off. It felt so good to still be lying in bed at seven in the morning, listening to the gibbons' duet, a melodious back-and-forth, starting with low whoops and rising up the scale to a delightful musical crescendo before dancing back down. On sunny mornings, the gibbons would greet the dawn with their call, but on mornings like today, they, too, stayed in bed late, and sometimes their songs took on a more mournful tone. The children would usually join in, as they learned from their parents how to sing beauty into the world.

Despite my joy at having a day off, I was worried. By then, Cam and I had been alone at camp for almost four weeks—ever since Jennifer and Alex left to go to the hospital. One unlucky day, Jennifer leaped into the river and immediately screamed in pain as her foot landed on a rock at an odd angle. She spent a few days hobbling about on a set of crutches hand-carved by Tadyn and Jono, but the pain was only getting worse. We used the shortwave radio to contact Serukam hospital, taking advantage of the weekly health slot when people across West Kalimantan could radio their questions. These sessions were also a good way to get the gossip from the few missionaries and anthropologists who had shortwave radios—but if you wanted to ask the nurse a question, you usually had to wait in line.

"Charlie Bravo Papa calling Sierra Echo Romeo. Please come in." After a few repeats, a scratchy midwestern accent would respond, "Charlie Bravo Papa, go ahead. This is Sierra Echo Romeo. How can we help you?"

We explained about Jennifer's foot, and the nurse brought Dr. Wendell Geary onto the call. Dr. Geary asked me to do various examination moves, and we relayed the results. The unfortunate diagnosis was that her foot was probably broken and that Jennifer should make the three-day trek to the hospital to get an x-ray and a cast. Since she couldn't come back to the forest in a cast, she'd have to stay there until the bone was healed. Jennifer was devastated, but she really didn't have a choice. Alex, incredibly kind as always, said he was happy to go with her. I knew she would be well cared for, given Alex's almost uncannily persistent good humor and endless energy.

Alex was originally from the Maluku Islands in eastern Indonesia— the history-warping Spice Islands, home to spices that were incredibly prized by European powers because the strong smells were thought to protect from the plague. Indonesians do not frequently travel outside their country, but within Indonesia, there is a strong cultural tradition of *merantau,* which loosely translates to "seeking one's fortune away from home." Thus he understood what it was like not to have family around. Having both Alex and Jennifer gone for so long was increasingly difficult. Alex's enthusiasm for learning and his constant jokes were a much-needed counterpoint to our Western tendency toward stress and anxiety.

And Jennifer provided a reality check to my own spiritual angst, as well as a buffer against the emotional waves that surged between Cam and me.

Now Cam and I were left alone: candlelit dinner every night, bathing in the clear water, trips to the waterfall, and spending every morning and evening alone together (except for the fifteen minutes I spent organizing the day's activities for the research assistants). It was all a bit too romantic, and we were having a very hard time sticking to our resolution not to get involved. The previous night after walking back together to our huts, we had hugged each other slightly longer than quite appropriate at the turnoff to his hut. We had switched off our flashlights and afterward we just stood in the dark with the sounds of the forest all around us. He was a dark silhouette towering over me, but I had the distinct impression that he was fighting not to reach out and kiss me. His deep voice cracked slightly as he got out a "Good night" and turned abruptly away. I had shared my history with him, hoping it would warn him away, but somehow he had been incredibly compassionate and kind and still seemed to see me in a positive light. Nevertheless, we were both averse to starting a relationship, and I suspected part of his internal struggle arose from his conservative Protestant upbringing. A rain day meant that we were going to be alone together *all day*, and this was just going to make things harder for both of us.

I finally decided to get up. I tucked up the mosquito net and shook out the thin kapok-stuffed pad I slept on, draping it over the string stretched across my hut. I nervously glanced up into the rafters to make sure there were no more snakes. The previous week, I was jolted wide awake one night by the soft thud of something landing on the floor next to my head. When I grabbed my flashlight, I found nothing—but the following morning there were two vipers in the rafters.

I kept on the magenta tank top I had slept in and pulled on some shorts. The best way to keep clothes from developing "the funk" was, crazily enough, to wear them as much as possible, as the body's heat would keep them semidry. Otherwise, clothes just imbibed moisture in the supersaturated air, creating the perfect medium for bacteria and

fungus. Already, almost all my clothes were rotting away. A precious letter from my sister said, "You're the only woman I know whose under-
wear rots off her. I just change mine when I get tired of them." My field
clothes had mud smears, snags, bloodstains from leeches, and rotting
edges. I left them hanging for a well-earned day of rest.

I slipped on my flip-flops and grabbed my flashlight, in case I didn't
come back until after dark. Using one of the rusting, semi-broken um-
brellas, I headed along the trail downstream to the main camp. As I
walked, my mind followed its familiar ruts, running over the questions
that had been plaguing me for months. I was increasingly aware of just
how ridiculously blessed my life was. One afternoon, one of the field
assistants, Tadyn, had come up to the main camp to get help with a
large cut on his hand. With his huge muscles and his brilliant forest
skills. I couldn't imagine him being afraid of anything, but when he
unwrapped the dirty cloth around the wound to show me, he looked
up with sheer terror in his eyes. The cut at the base of his right thumb
was pretty deep—but it wasn't life-threatening. What was the big deal?
But then I suddenly realized that he was right to be shaking with fear.
His very life *was* at stake. And so would anyone's, if they had never
had a tetanus shot, did not understand the germ theory of disease, had
no access to antibiotics, and were dependent on the use of their right
hand to financially survive. I sat down, deeply distressed. Not by the
wound but by the realization of what it meant to live in a place like this
and how lucky I had been in life.

Using a well-worn copy of *Where There Is No Doctor* by David Werner, I had learned how to clean a wound and make butterfly stitches
using tape. Tadyn's thumb was so muscular that the wound had popped
open like an overripe grape, but I was able to pull the edges of the cut
together, and with the tape, I made a nice, neat line out of the gash.
The book also suggested the best choices and dosage of antibiotics,
which I obtained by rummaging through our first aid kit. It had been
surprisingly easy and strangely satisfying. It had felt good to see calm
return to Tadyn's eyes.

"Thank you, Kinari. You're good at that. It barely hurt at all. Maybe you should become a doctor?" Tadyn winked.

Smiling at him, I thought, *Yes, maybe I really should.* The seed that was first planted at Serukam was steadily growing.

As I walked along the path, I wondered: If you were as lucky as I was, and you just happened to be born into the richest country in the world, at a time when feminism had already made some inroads, and you just happened to have a pretty good brain, did you have a responsibility to share that good fortune? I wasn't an inherently better person—if anything, the opposite, and I was still tortured by a feeling of having a rotten core—but somehow I had received these blessings.

Cam had told me in one of our philosophical discussions, "While there may be no verification of God's existence, there can be *validation.*" I had given that a lot of thought. Being certain that God did not exist might set up a barrier that could make it hard to experience her/ him/them/it. But if one is open to the possibility of Something Else, it might become likelier that the position would be validated. Maybe it depends on where you are willing to shine your flashlight. If you refused to look in the direction of the divine, you might not ever see anything. In other words, *faith* might be required—at least, enough faith to look in that direction.

I wasn't accustomed to these kinds of theological arguments, and I wasn't sure I really cared why the world was the way it was, since the answers were ultimately unknowable. However, I was getting to the point where I could no longer be certain God/Spirit/Other did *not* exist, and just that slight change had shaken me.

Walking in the rain, I looked down and saw a huge toad, just off the trail near my feet. I squatted down to look at this grapefruit-size brown amphibian and was amazed to see that it was covered in tiny mosquitoes. I then noticed a slight movement on a nearby bush and realized that what looked like a leaf was really an insect mimicking one. And the mimic was perfect, absolutely perfect. Even down to the corner that looked like it had been gnawed by a caterpillar.

I sat back on my heels and looked out over the river, the rain gently washing everything clean. I felt like weeping for the beauty of it: the mist rising into the canopy, the twist of a vine, the lush forest arching over the clear river. No human creation could even begin to touch this splendor. The Taoists thought that natural beauty itself was an argument for God. And if God did exist, wouldn't all people and all religions be searching after the same inherent truth? A truth that none of us could possibly grasp, just as I would never be able to describe the exquisiteness of this view. Each religion probably gets some aspects right and some wrong— each one just a single arrow, clumsily aimed in the direction of a Something that can never be described. Living surrounded by the incredible diversity of the forest made me aware of how much we don't know.

I had been reading a book my mother sent me on chaos theory, and I found it exciting and uncomfortable to think how interconnected everything is. The proverbial butterfly, or that leaf insect in front of me, could flutter its wings and affect storms in the United States. What did *that* mean for morally right and wrong choices? Could one ever know the implications of an action? What were the ripple effects of people here in Borneo not having health care? If they had to illegally log to pay for medical services, was that affecting the health and well-being of the whole world?

I shook myself from these abstract thoughts and stood up. I couldn't keep putting off going to the main camp. Cam was driving me crazy— literally. I didn't know how I was going to spend the whole day with him. The more time I spent with Cam, the clearer it was how much I cared about him. I'm sure back home I never would have given him the time to get to know him so deeply—if I would have even crossed paths with him. He was not particularly gregarious, and the first impression he often gave was of gruffness, but here, with our long, intimate conversations that bared his kind soul, I found myself unable to resist. I was falling in love, and I feared the same was happening to him. I did not want to hurt him the way I had hurt nearly every one of my previous partners. Part of my past behavior may have been a fear of true intimacy and part of it a fear of being controlled. I was determined to become a better person but wasn't sure I had the strength to do so.

Away from the forest, I knew he would find someone more suitable who would make him happy. He was tall, handsome, well educated, funny, brilliant, thoughtful, and generous to a fault. On the other hand, he did have a short temper and could lapse into brooding misanthropy, but I didn't doubt for a second that he would be snatched up by some lucky woman. That, I told myself, was what I wanted for him.

A few weeks before, Cam told me how appropriate he found my last name of Thomas. Thomas was the disciple in the Bible who refused to believe even when standing in front of the resurrected Jesus until he had put his finger in the wounds. In front of Cam, I maintained my stubborn certainty in death, my conviction that everything was explainable, and my unwillingness to consider anything beyond what I could literally dig my fingers into. This conviction was less strong than I let on; but he was right, I fully empathized with the doubting disciple.

Something about the way Cam had talked about my last name made me think he would like to change it in more ways than one. And his last name, Webb, was also symbolic of how he was trying to get me to see the world—as utterly interconnected, both in an ecological and in a philosophical sense. But it was also clear to me that our growing relationship was sheer torture to him. He was drawn to me and repelled by me, and in this microcosmic human society, it was completely impossible to avoid each other.

Our task that morning was to collate a trove of dried specimens, spreading them out on the floor in their folded squares of newspaper. We set to work, awkwardly avoiding being too close, until we accidentally reached for the same specimen and our hands touched. Cam jerked his hand back as though shocked. He leaned away from me, but then that wasn't far enough, and he stood and walked backward. "Kinari, I can't do this anymore. It's too hard for me. Would you pray with me? I know you don't believe, but *I do*—and we both need help to make the right choices."

Still sitting cross-legged on the floor, I put my head in my hands.

Had I gotten to the point where I could believe, for just one minute? After all, it was *possible* the Divine existed. And it would mean something to Cam. I knew I wasn't ready to talk yet, but maybe I would be willing to listen. Looking up slowly, I said, "Okay, I will pray with you. I think I can believe long enough for that."

We walked to the table and sat opposite each other on the wooden benches. We folded our hands and bowed our heads, in the semiotics of "prayer" within our own cultural tradition. My desire to help us both find some peace and strength was enough to be willing to try. I felt a softening and opening in my chest as I gave in for a second. It was like the image in the Sistine Chapel, where Michelangelo represents the human reach for the Divine as merely the limpest gesture—a finger barely lifted toward some possible Other.

And then, a nanosecond later, electricity jolted down my spine: my whole body felt flung back as every particle burst awake, a maximally outstretched energy zinging through my being. My head was up, my eyes were open, tears streamed down my face; awe and amazement filled me, speech was impossible. The entire world transformed. The forest through the open sides of the station was crystal clear, exquisite in its complexity and wholeness, as though I had only ever seen it through distorted glass. The difference between looking at DNA and knowing the love of a human being. My whole life, I had been seeing just a frog, or an insect, or even just rain in a scene, and I had missed the whole. *How could I have not seen it before? I had been surrounded by this truth my whole life but had been completely blind to it.*

Me, Cam, the forest, the whole universe, part and parcel of all. *Sacred.* All perfect oneness. Complete love and beauty but also extreme power. Separate yet also me. No boundaries. Not a soft malleable Other that I could bend to my irrelevant will but rather the fullness of isness and the maximum of potential. All the energy of the universe flowing, living, glowing, humming, delighting in existence.

Cam looked at me with moist eyes and a stunned smile, confused and overwhelmed to see my shaking body, my tears, and my mouth hanging open and mutely moving. I could see love in his eyes, as well as awe. I stood up, stumbling, completely unable to talk. I left for the forest, not knowing where I was going, and Cam watched me go.

I wandered, seeing for the first time. The forest was life and color and vibrancy. Totally alive. Way beyond anything I could ever understand, from the smallest molecular particles to the most complex rippling interactions—even on the scale of the universe. I passed my hut and kept going. Hearing for the first time. Feeling every slight drop of rain on my skin. Awake. Awake as I had never been before.

I came to the river. Now the rain was stopping. I didn't know why, but I knew I had to get into the river. I shed the filthy layers of clothing and walked into the pool. I submerged myself, feeling the water flow over my body and seeming to flow *through* it, aligning my molecules. I stood and was only mildly surprised to see light streaming down. Further validation.

I walked out of the water toward the sand beach that formed at the bend of the river. Then the walls slammed shut, and I collapsed on the sand. *What am I thinking? Has my brain gone crazy? I'm a scientist! This must be some kind of weird hallucination or delusion. And what am I doing lying here? There might be ticks on this sand from the wild pigs.*

Then another part of me asked myself: *If you're truly a scientist, are you going to reject the data because it doesn't fit your hypothesis? That's not the way science is supposed to work. You're supposed to mold your hypothesis to the data, not the other way around.*

But how can I trust what I just experienced? This is crazy; I can't possibly believe this. I don't feel insane, but maybe I am. Maybe I'm having an LSD flashback? Unlikely—I haven't tripped in more than five years. Maybe . . . maybe . . . maybe . . . I am actually seeing correctly for the first time? Maybe I finally listened, and I just didn't like what I heard.

No, that can't possibly be it. Because if this is real, it will change every-thing.

(6)

Deer Paths

NEW HAVEN, CONNECTICUT:

1994–2001

After that transcendent experience in the forest, I struggled for a long time whether to accept what had happened to me as a revelation of truth. Eventually, I knew that I had no choice: I would follow the evidence. I found myself having to rebuild my understanding of existence and myself. In the process of that opening, other impossibilities also began to feel possible—including Cam. We each softened to the other and to ourselves—accepting the truth of how we felt. Though I'm pretty sure this isn't what he had in mind when he asked me to pray with him, it ended up being one of the consequences.

Shortly after that profound experience, we were no longer alone, and camp was full of bustle. My father, his girlfriend, Mark Leighton from Harvard, Cam's advisor from Dartmouth, Jennifer, and Alex all arrived at the same time. Cam would be traveling out with all the visitors to teach classes the next semester, so there was little time for us to talk. But during this time just before he would be leaving, he got sick for a few days (likely something brought in by the visitors), and I took it as an excuse to bring him food each evening. One night, I lay down next to him on the bed as we talked. Silence descended, though, and we just stared at each other. "I want to kiss you so badly, but I know I shouldn't," Cam finally said. I thought about it for a while, and then replied, "Well, I could always kiss you, and you don't have to kiss me back." Let's just say, that plan didn't work. We melted into each other and were both left breathless.

After he left, letters came and went with every boat trip, each bringing us closer together. His poetic prose further enchanted me. In our letters, we shared more vulnerably about our souls and our dreams for our future lives. This brought further intimacy, but the idea of marriage, which he strongly hinted at, still terrified me.

During those six months, I struggled not only with the question of long-term commitment and my capacity for it but also with religion. If God did exist, clearly all humans through all time would be seeking to know the same truth, which would mean there would be wisdom in all faiths. Should I then align with a religious tradition? A few months after my experience, I wrote a letter to a close Quaker friend from college, Julia Riseman, who had founded Reed's volunteer program. I told her about what had happened to me and shared my anguish about whether I should choose a faith. On the next boat—a month later, way too soon for the minimum three-month turnaround—I got a letter from Julia saying that their second daughter had been born, *and they had named her Kinari.* As our messages of joy crossed once again, hers came with a Quaker book called *Notes and Queries* that offered questions about faith and Quaker responses.

This book proved pivotal for me. Here was a group of people who were willing to die for social justice, were pacifists, believed in the equality and preciousness of all humans before God, and even practiced meditation. But still I believed that there should be no divisions among religions, and that freed me to pull in wisdom from other traditions while staying grounded in a community of faith.

After my initial intense experience in the forest, I had been able to feel the oneness of everything easily and often, but over time, that faded, and I found myself, at times, worrying that God did not exist. It's amazing how fickle humans can be—or maybe I should just claim this for myself. Kindly, though, I was often given affirmations and sometimes clear messages. One of these was when, meditating during my last months in the forest, I heard simply: "You *will* marry Cam." I wasn't sure if that was an order or a prediction, but it did help relieve my anxiety about the choice. Even though we had not explicitly discussed it yet, it was clear that was the direction we were headed.

During a short overlap in Indonesia as Cam traveled back into the forest and I left for my last year of college, we escaped for a week to a tropical island. One evening, I climbed down from the porch of our thatched beach hut looking over aquamarine water. At the edge of the glistening white sand, I found a little bush and broke off a flexible twig. I clipped it to the right size and then without a word offered it to him, showing him the size of my ring finger. His grin silenced both our fears. With thick pure gold rings cut from the same Bornean nugget and stamped with leaves representing the forest where we met, we officially proposed to each other the next Christmas when he unexpectedly came to visit me at Reed. I felt I was too young to be marrying, but fate, it seemed, had set the course. Weeks after I completed my undergraduate degree, we were married on a mesa top in New Mexico, surrounded by friends, family, and beauty. Cam wore cowboy boots and a striped silk vest that matched the colors of my bouquet. The sky itself felt like a guest, and my favorite birds from childhood joined the celebration overhead. Some might not consider turkey vultures gliding above a blessing, but I had a baby name for them ("chakoos") and loved the way they floated on the air without apparent effort.

I returned to Indonesia with Cam for his final field season the winter after the wedding. I was struggling with choosing between medicine and conservation as a career, and this trip was an opportunity for me to explore both options. I volunteered at Serukam hospital and traveled to remote communities with their village health program. Diving into their data on growth and mortality, I found that the percentage of children dying before age five had been cut in half—from the initial jaw-dropping 25 percent to a (still-high) 12 percent. That was a lesson in the powerful effect of simple interventions like immunizations, supplemental nutrition, and health education—*and* in having the data to show the impact. I also joined Cam on research trips in three different mountain rainforests. Sitting on a beach on the tiny offshore island of Karimata, where the forest came down to the sand and coral reefs ringed the shore, I made long lists of the pros and cons of going into each field, finding it impossible to choose between the two. The problem was, they were both critically important.

How could I decide between healing people or caring for nature? It was a false choice anyway; the two were utterly intertwined. I had seen the connections clearly in my travels in Kalimantan. Humans could not survive without a healthy ecosystem to live in; and in some areas, access to health care is so limited that people are driven to destroy their future to obtain it. I began to feel a strong sense that I was supposed to work on *both* human and environmental health. The Blackfeet cosmology understood that both were needed for community actualization and sustainability. I decided to apply for medical school but with the goal of using those skills to work toward thriving people and thriving ecosystems.

I found the whole medical school application process wrenching, because I didn't believe I would be accepted anywhere. Without Cam's steady support and kindness, I wouldn't have had the gumption to do it. When I told interviewers that I planned to return to Indonesia, they seemed to think I was crazy—or maybe just delusional. But somehow, I did manage to get in, and I found myself living in New Haven, Connecticut, in 1998, attending Yale's medical school.

Slowly, at Yale, my belief in myself grew. One of the most useful things I was learning was that the distribution of human intelligence isn't actually very broad. Ninety percent of us fall in the middle of the curve, and none of us are all that different from each other. I had grown up being told that I wasn't very intelligent, so discovering that we are all basically the same was hugely empowering. I even did well academically, earning praise in my clinical rotations and awards. One of my professors made me blush when he told a large group that, when he was old, he hoped I would be his doctor—because I was so good at diagnosis.

During the first year, Cam was at Harvard doing a postdoc, which meant we saw each other only on the weekends. He became frustrated and angry with the two-and-a-half-hour train rides to New Haven, and it wasn't working for me either, because I badly needed his support. We couldn't see a solution to this problem. But then, completely unexpectedly, Cam's postdoc advisor decided to take a job at Yale—and he asked Cam if he would be willing to transfer with him. I saw this as another miraculous affirmation we were on the right path.

Cam's move to New Haven meant quite literal provisions for me, because when he was away, I just ate cereal. During a surgical rotation, I began my days at 4:00 a.m. and finished at 8:00 or 9:00 p.m. I discovered that the most efficient way to get sleep and food was to come home, run the bath, make a bowl of cereal, start soaking in the tub, scarf down my corn flakes, wash, drain the water, and get in bed. Boom! Eating and bathing accomplished in just fifteen minutes. When he was home, though, a hot full meal often awaited me.

It was also helpful to be married to someone outside of the medical world. Medical school was a mix of amazing teachers, incredible comrades in arms, and fascinating things to learn, along with some examples of exactly what I didn't want to foster in my own life. The positive aspects included increasing empathy, evaluating evidence, realizing the power of a team, and learning to truly listen. Listening isn't actually an intuitive skill, and the data shows most doctors aren't very good at it, with only eighteen seconds going by before they interrupt. Yet listening well can literally be the difference between life and death. I learned to pay attention to both words and body language; mirror patients' emotions; look in their eyes; reflect back key words to make sure I understood correctly; and make small noises like *emmm* and *ahh*. Small things like orienting your body toward the speaker and having a posture of openness (no crossed arms) also encourages people to share. Really wanting to know what they had to say was also critical. As one of my professors said, "The secret of caring is *caring.*"

Then there were all the things that I wanted to actively counter: depersonalization, sexism, near-medieval levels of hierarchy, and obscuring simple ideas with complicated language. My belief in the importance of data was also affirmed, but sometimes by its absence. My scientist husband was incredulous to learn about the new focus on "evidence-based medicine." "What on earth had they been doing before?"

Somehow, a cluster of doctors standing around a half-naked patient and discussing "the case"—while barely acknowledging the individual as more than a disease—had become disturbingly un-disturbing. Thankfully, I later did a rotation in the Navajo Nation. There I was appropriately chastised by a wonderful family medicine doctor, after I asked him a question in front of a patient in exactly the way every Yale medical student acted. He yanked me out of the room and yelled at me, "She is a human being!" His anger at my disrespect made me realize how far I had fallen in just three years.

I wasn't alone in being frustrated with the medical culture, and we had many amazing teachers promoting a more "patient-centered" approach. One of my favorite professors was Dr. Thomas Duffy, who was legendary for his diagnostic skills as well as his compassion for his patients. Simply by looking at patients' fingernails, he could often figure out the illnesses that plagued them. I sought out these remarkable mentors, soaking up their wisdom and even becoming friends with many of them.

I was surprised to discover what a huge role sexism still played in medical school. More than half of my cohort at Yale were women, but few of our professors were. At an interview, I was asked how I felt about being a woman going into "a man's profession." I thought that was absurd—until I started attending classes and saw the truth of it. Even before I started, I encountered resistance. When my mother-in-law learned of the plan, she exclaimed, "But what about Cam's career?" I responded, "What about mine?" Her response was telling: "That's different. *You* are a woman." Eventually, Cam's parents had come around, helped likely by my assistance with some of their own medical concerns.

One of our cadavers was even the body of one of the first women ever admitted to Yale medical school—in other words, it hadn't been very long since women had been allowed in. I was also shocked to discover that almost all medicines have been tested only on 150-pound white men. That means if you are small, you are likely being overdosed, and if you are large, you are probably not getting enough medicine. And then, if you are female, or not of European descent—who knows whether these medicines will work the same way for you. They might even make you worse.

In other ways there was also an incredible amount of racism in the entire way medicine was structured and very little discussion about it. I knew it would be a struggle to counter my own internal biases, but I wanted to be respectful of everyone and aware of the ways that "care" could lead to harm in the medical system. This was true not only in the doctor-patient encounter but also in the unjust way society was structured—which affected who got which illnesses and often how they were treated.

I was grateful to have professors like Dr. Woody Lee, a cardiologist who taught my small group on the doctor-patient encounter. Dr. Lee's experience of reaching the very top of his profession, despite racism against an African American man, may have been part of the reason he was so brilliant at helping us see the perspective of others no matter what background they came from. We would spend hours with his heart transplant patients, listening to their wrenching life stories that often left us teary-eyed. It was such an incredible privilege to experience our patients' generosity as they shared their most intimate fears and deepest secrets. We could ask about anything—in fact, we were expected to—and in the process I was learning about the full scope of the human experience, from its worst pains to its highest transcendence.

All these experiences taught me critical lessons about how I wanted to practice medicine and how I wanted to be in the world. At the same time, Cam and I were learning more and more about the state of the planet, rising levels of CO_2, and overuse of resources. This journey began in the first place we lived together: the big communal farmhouse of Donella Meadows (Dana to her friends), the prophetic author of *The Limits to Growth.* That book and many dinnertime discussions with thinkers from all over the world, who would gather around her long oak table, had fostered even more clearly our understanding of how interdependent everything is. Harm to one, including to the environment, is harm to all.

While we already knew it to some degree from our experiences in Borneo, Dana showed us so clearly that the main road that most of society was galloping down headed straight for the steepest edge of the mesa. And like her, I had to believe that there were other ways down that, if taken slowly and carefully could lead to a valley where humans and ecosystems thrived. Certainly many Indigenous groups had found such valleys, so it had to be possible. As my friend Margaret Bourdeaux said, we weren't looking for the road less traveled, we were searching for a deer path. Finding this path on the other side of the ocean wouldn't be easy, but I knew that our only chance would be to follow the wisdom of the local communities.

(7)

Parachuting Rice

ACEH, INDONESIA:

FEBRUARY 2005

The first step in returning to Indonesia had been my choice of a specialty. In pursuing family medicine, where one cares for children and adults and does obstetric care, I got serious flak from some of my professors for not choosing something more prestigious and lucrative. But Nancy Angoff, our dean of students at Yale, got it exactly right: "What will you do, working in a remote place, if you don't know how to deliver a baby?" She also supported me when, after getting a spot in a family medicine residency, one of my professors at Yale crossed the street to confront me: "*Where* are you going again? *North Dakota?*"

"No, California."

"Ha! Same thing!" he snapped back, reflecting the Ivy League's bias against anywhere else. I didn't feel better when he added, "When you come to your senses, we'd still be willing to have you here."

So Cam and I moved across country for my training at Contra Costa Regional Medical Center near San Francisco. Cam hadn't wanted to move to New Haven, but when the time came to leave, he didn't want to go. Mostly, he didn't want to give up his position at Yale, but his irritation eased somewhat when his advisor allowed him to keep it and also be affiliated with a local university in California. And then, luckily, Cam ended up loving California, and, thankfully, caring for me during the crazy hours of residency.

As we approached my final year, another decision had to be made about next steps. Cam was starting to be actively headhunted for faculty positions, and he was feeling honored and tempted by them. I wanted to support him and also desperately wanted to fulfill my dream of returning to Indonesia. I had been looking for jobs with development organizations, but all of them scoffed at my idea that the health of humans and the health of ecosystems are interrelated. I struck out one after another until I was at a loss of who else to approach. It was hard for me to argue that we should move to Indonesia when neither of us had a job there, Cam had a good alternative, and we still had sizable debt from medical school.

The stress in our household kept growing, and neither of us knew what to do. Then in October, on one of my rare nights off, we decided to go to the movies. We saw *Beyond Borders,* in which Angelina Jolie plays a wealthy woman who drops everything to bring a convoy of food to a refugee camp in Ethiopia. We both cried through much of the movie. Afterward, we sat at a little sidewalk café in silence. Then Cam looked at me and said, "Okay, I'll withdraw my applications. I have no idea how we are going to pull this off, but let's try to make it happen." After dinner, we stopped while walking to the car and ended up just holding each other as more tears flowed. I felt like I had imagined Kristen the orangutan felt—knowing she'd chosen the right guy.

Then followed months of fear. We had decided to sail off into the unknown, but we had no provisions, and we knew we wouldn't make it without them. To top it off, in December 2004 when I was in the middle of the hardest rotation of family medicine training, working more than one hundred hours a week for six weeks straight, a tsunami hit the island of Sumatra. On Christmas Day, it swept across the entire Indian Ocean, killing almost two hundred thousand people.

I vaguely followed the details as I overheard snatches of conversation at nursing stations or caught a few words of television news while I examined a patient. I had no energy to process it. But every few days, when I had a few minutes at home, Cam would insist, "We have to go! We have *years* of experience organizing things in Indonesia. *We can help.*"

I kept telling him he was crazy, that I was doing my residency, it was impossible. But finally, he wore me down, and I went to go talk to my program director. To my surprise, Dr. Fish jumped on the idea. "Yes! This is what Contra Costa can do for tsunami relief! We can send our only doctor who speaks Indonesian. I'll rearrange your schedule and get people to cover your shifts. How about if we give you credit in emergency medicine? That should work well."

Dr. Fish quickly arranged everything, and the other staff were amazingly generous with their time and money. Then, absurdly, we could only find a position for me, not for Cam. But as soon as I got to Aceh, it was clear how desperately logisticians were needed who could speak Indonesian, and with my insistence, International Medical Corps (IMC) sent him a ticket.

Two weeks later, I was having a fight with the IMC staff.

"What do you mean, Dr. Irene will have to stay with the cleaning staff?"

The volunteers and staff were assembled in one of the fanciest houses in the town of Lamno, Aceh, just beyond the horrific destruction the tsunami left in its wake. I had been woken up three times the previous night by terrified people convinced that a family member was dying, though each time it turned out to be a (completely understandable) panic attack. As the only doctor in town who could speak Indonesian, I was the preferred doctor to call upon. My patience was worn thin at this point. I needed someone to share the night calls and listen to the heartbreaking stories, so I had begged IMC to allow me to invite an

Indigenous doctor and dear friend, Dr. Irene, to come and help. I knew her from Serukam hospital, where I had returned during both medical school and residency.

I was worried that if I didn't get assistance, I wouldn't be able to give my patients the only gift I had: full witnessing. There is a hyper-expressive phase to trauma, and these people were in it. Every single story was, truly, the worst story I had ever heard. Mothers who had their children ripped from their arms—"I couldn't hold on to them, *I couldn't!*" A child stuck in a tree for three days, a crocodile lurking at its base. A woman selling her wedding ring for food, after huddling for days in the rain—afraid to come down from a hill, knowing she would find her parents dead at the bottom. Or the young man who sobbed as he told me about outrunning his grandmother who was caught by the wave just behind him. They went on and on, one after another. They all desperately needed to tell their stories and have them heard.

"I'm sorry," I was informed by the medical chief. "IMC's policy is definite on this: *nationals* have to live separately."

I was flabbergasted by his racism. "Fine, then I will move to the other house as well! This shouldn't be about *us* helping *them* but about working together. She is a doctor, just like you, and she deserves respect. She is also one of the most wonderful human beings I have ever met."

He scowled at me but eventually relented.

Shortly after this conversation, people began filing in for a coordination meeting for all the nongovernmental organizations (NGOs) working in the area. Everyone settled into a big circle of plastic chairs.

Following a few opening remarks, the area head of Doctors Without Borders reported, "The governor in Banda Aceh has given us control of the government clinic building beginning after the Pakistani military leave, in three weeks' time."

The IMC logistics head was indignant. "Well, the Indonesian military commander here said he would give it to *us*! And since this is still a war zone, *he* has control, not the governor."

Mercy Corps had a different view. "We put in a petition to the department of health, who *should* have jurisdiction, and we were told that the decision had not yet been made." The territorial jousting continued.

Then the discussion turned to Aceh's separatist movement, known as GAM. The Indonesian military was demanding that the NGOs not provide medical care to GAM members. The military was also complaining that many of the NGOs were displaying NO GUNS signs on their cars, which meant that military personnel could not hitch rides with them. This topic was dispensed with rather quickly, as it was obvious that no one cared what the Indonesian military thought. I was amazed at this arrogant, colonial attitude; but in truth, it was unclear exactly who was in charge. Thousands of people from all over the world had flooded into the region, effectively stopping the separatist war by their presence. In this strange situation, the military seemed to be scrambling to assert some control—and failing.

The next topic was housing costs. A Red Cross / Red Crescent volunteer piped up, "Will you please start bargaining a little for the houses you rent? The prices are doubling every week, and they're already approaching New York City rents!"

One of the nurses volunteering with IMC spoke up. "No one should order any more malaria medication. We have a room so full of medicine, we could treat the entire population multiple times, and so far there hasn't been a single case of malaria."

I was pleased to hear her say this, since she had seemed generally way out of her depth. IMC had brought in many volunteers who had never been outside the United States. The rotting bodies, the horror of the destruction, and the chaos of the displaced people—all overlaid onto the existing war—was overwhelming for us all, but some of the American volunteers were barely holding it together.

And then there were the professional relief workers, some of whom flew in from assignments in Darfur. I knew there was a certain type of aid worker who was actually seeking adrenaline, and I knew that these were often people who had grown up in chaotic homes or societal contexts and who felt normal only in crises. However, many of them coped with the stress by drinking heavily and playing loud music. While I could empathize with their need to blow off steam, I could not excuse their lack of respect. This was a conservative Islamic society, and *everyone* was in mourning for lost friends or family members.

For the people of Aceh, the level of stress was almost unimaginable. Our driver said something poetic. "The night of the tsunami, our pulses were already very fast. There was shelling between GAM and the military, and we were all terrified. The wave came in with the flare of rockets—and the beating of hearts." The war had been put on hold—sort of—but the tensions were present everywhere one looked. The horror of the conflict was epitomized by a man who told me that the tsunami would actually be worth it if it stopped the war.

My attention was pulled back to the NGO meeting by a rising discomfort in the group. A gentle, Indonesian-speaking volunteer from a small nonprofit called Millennium had thrown an unintentional bomb into the meeting.

"We've been talking to the communities, and they say they are very grateful for all the medical care and the help with food and shelter, but what they need more than anything else is help clearing the rice fields of debris so they can start planting again." He explained that the farmers were worried that if they didn't plant soon, in three months, there would be no food for anyone to eat. They would rather continue to live in tents and not have medical care than not be able to plant. He said he would try to get Millennium to help, but they were small, and he would be grateful for help from some of the larger groups with more resources.

Stunned silence. I almost wanted to laugh. *He actually talked to the people here and asked them what they needed?* That behavior was incredibly rare. The standard belief seemed to be: *We know what is best for you! We are the experts.* That is why we had an entire storeroom full of malaria medications for a region with no malaria—and no sanitary napkins for women, even two months after the tsunami.

Not surprisingly, there were declining murmurs from all the big NGOs. They were sorry, but despite raising many millions of dollars to help after the tsunami, clearing rice fields wasn't in their brief for disaster relief, only distributing bags of rice (probably imported from the United States).

As the meeting moved on to other topics, one of the few Indonesians present got up and stumbled out of the room. A few moments later, he poked his head back in and gestured wildly to me. I slipped out to join him, and he asked me in an anxious voice whether I would examine his heart. I went and got my stethoscope and found his heart racing and skipping beats.

"Do you think it could have anything to do with the methamphetamines I've been taking?"

I sighed deeply. "Yes, I'm afraid, probably so."

He was one of the few Indonesian senior members of an aid organization, and I was sad to see that either they had chosen an addict or he was getting so little support that he had turned to drugs to cope.

"Come on. I'll walk you over to the Pakistanis. We can put you in a bed and monitor your heart." I grabbed a scarf from my room to cover my hair and steered him toward the door.

We walked down the middle of the narrow main road, since there were so few vehicles. A man intercepted us.

"You're the American doctor who speaks Indonesian, aren't you?" I nodded, and he continued in Indonesian, "I've been wanting to tell you how grateful we are to the Americans. We are happy that everyone has helped us, but we are especially grateful to the Americans because you fed us when we had no food. Three days after the tsunami, when we were starting to starve, American military planes flew over and dropped food for us with parachutes. Please tell your countrymen how grateful we are!"

"I will." I smiled for the first time in many days. At least that American rice had been useful at the right moment.

When we arrived at the *puskesmas* (community health center), I felt the same relief I felt every time I entered; this was the only place that was run smoothly and efficiently. The Pakistani military had done a brilliant job of turning this little clinic into a field hospital. I got my patient settled with one of the internists and then checked on a few of the female patients I had been helping with. (All the Pakistani doctors were men, and some of their patients wanted to be cared for by a woman.)

Before going back, I decided to poke my head out the back door to see if one of my favorite people in the world had time to talk. Colonel Muhammed Daud was sitting at a table under a military-style tent with its sides rolled up. A huge pile of papers sat in front of him, and a secretary stood stiffly by his side, accepting each paper after it had been signed. I was just pulling my head back when Colonel Daud looked up and saw me. "Ah, Kinari. Please come have a cup of chai. I'm sick of all this paperwork anyway. You are the perfect distraction."

His lilting accent, exquisite manners, and calm were always so welcome within the swirl of chaos. With a wave of the colonel's hand, his secretary gathered up the paperwork and went to fetch two cups of hot chai. *Just what I needed.* I settled into one of the military green canvas folding chairs and took a deep breath.

"How was the coordination meeting?" he asked, displaying his customary omniscience.

I rolled my eyes.

"Ah, a standard one, then." He laughed, making me smile. "It is the same with relief efforts all over the world: territorial fighting, wasting money, and—as an occasional side effect—actually helping people. Remember the Mercy Ships helicopter that 'miraculously' rescued the baby's life you had already cured—and then didn't have room for the mother because of the film crew? That's sadly typical."

Shaking my head, I confessed, "Colonel Daud, I don't know what to do. I am feeling so frustrated with all these NGOs, and it all comes on top of worries about my own life."

"Please tell me. I have already told you about my woes. I should not be a good friend if I did not listen as well." Colonel Daud had told me about the son whom he loved more than life itself, who was in constant pain from a malformed spine. When they went to the United States for the first of many surgeries, he told me how kind and generous the nurses had been to his son. That experience had strengthened his resolve to resist the Pakistani extremists who would denounce Americans. *"Are we not all human?"* were his words then.

I told him about how I had tried to get jobs with many of these same organizations, but now I was grateful they wouldn't hire me, since none of them seemed to have any desire to actually listen to the communities, and they were so dysfunctional. "Being here in Aceh makes Cam and me both feel that our decision to return to Indonesia is right, but I have no idea how it is going to work out." I looked down and paused. "All I know for sure is that it has to be about actually honoring the people you are working with. The key to doing the right thing will *always* be listening. I just have to have faith that the answer will come. But simply trusting that somehow it will work out is so hard."

"In Islam, we would say, 'God is great,' and believe that God will open the doors if you are looking for the right path."

I noticed a number of people discreetly waiting for the colonel's time, so I thanked him for the tea and kind words and promised to come back the next day. As I left the *puskesmas* and walked back down the central street, my thoughts were swirling. It felt so precarious to have decided to move to Indonesia without having any way to survive, and now we had missed the deadlines for the academic cycle. *What were we thinking?*

As I walked back to the IMC office, I heard a jungle call, and when I looked behind me, I saw Cam jogging to catch up with me.

"Kinari, you are not going to believe this! I have the most amazing news! I was down at the soccer field, helping with a shipment of mosquito nets. You know, for the malaria we don't have. But the helicopter pilot let me borrow his satellite phone, and I was able to check email."

"What did you find out?" His infectious joy made me smile.

"I got an email from Bob Cook. Do you remember him? I worked under him at the Arnold Arboretum when I was at Harvard. He heard about how we want to come out here full-time and that you want to do a program that would combine health and conservation. He wants to support *both* of us, so he said he can offer me a job through Harvard but based here in Indonesia! I'll get a salary to do whatever research I want!"

At first, I could only stare at him, dumbfounded, but then I started jumping up and down, almost screaming. Colonel Daud had been right—miraculous provisioning for the journey had been provided. Rice had arrived in a parachute from the air just when we'd needed it.

8

Leaf Monkeys and Butterflies

O ne Sunday, after returning to California from Aceh, I sat in our little enclosed garden, surrounded by potted plants thriving in the sun. I was talking on the phone with my good friend from college, Julia Riseman, who had named her daughter after me during my first year in the forest. Julia was upset to hear my stories of how the nonprofits behaved after the tsunami, and she agreed that there was no way I could go to work for any of them.

"But, Julia, what am I going to do?"

Julia paused and then said, "Well, maybe we could start our own nonprofit. One based on the principle of listening to communities."

"That's possible? Actual humans do that sort of thing?"

The more we talked about it, the more certain Julia was that we could pull it off. Then, over the following months, my friends came out of the woodwork to help. First, we had to come up with a name, and it was appropriately birthed at the home of a friend whose baby I had helped deliver. Doulaed by good farmers' market organic food, wine, and a giant thesaurus, Health In Harmony came into being. We liked the way it played off the phrase *in harmony with nature*. We bought the website URL that same night.

But starting a nonprofit requires more than an idea, a name, and a URL; it requires an immense amount of paperwork, and neither Julia nor I knew how to do it. The solution came from an unexpected quarter. Before leaving for Indonesia, I made a trip across the United States to gather support and seek wisdom from people who had experience working in underserved areas. In New Haven, I was sitting in the kitchen of two good friends, Mark Totten and Kristen Rinehart-Totten. Mark, who was then about to start his last year of law school at Yale, offered to take on all the legal paperwork as his senior thesis project. With advice from his professors, he did everything so perfectly that, instead of the standard six months, we got nonprofit approval in just three weeks. Mark and Kristin then joined the board of Health In Harmony, along with Julia and my close friend from medical school, Anna Hallemeier, and the amazingly supportive dean of students at Yale, Nancy Angoff.

One of my colleagues in residency, Ann Lockhart, also joined. Despite knowing nothing about accounting, she became our treasurer and did a brilliant job managing all the donations and finances. The highlight of her week, she said, was to stop in and pick up the checks at our post office box. Julia, Mark, and Ann were each working at least twenty hours a week for Health In Harmony, and tens of other friends were also putting in many hours—all for free.

I was surprised there were people willing to invest their time and money in the project when it was still only an idea. At one fundraising party my dad threw before I left, I overheard a woman remark, "It's great that Kinari has this big vision, but it's just a pie-in-the-sky idea." The man she was talking to responded, "Well, she graduated with honors from Yale Medical School, and that takes intelligence and persistence, so I'm betting she can do it." I hoped he was right—but maybe her skeptical view was closer to the truth. My mother used to say, "Well, you're not very smart, but at least you work hard." Hopefully, that would be enough to carry us through.

WAS IT TOO LATE? WAS I on a fool's errand?

I spent my first year in Indonesia traveling across the country looking for the right location to begin. Dr. Irene, my friend from Serukam hospital who had come to help after the tsunami, often traveled with me. Irene's parents were Indigenous Dayaks from Central Kalimantan, and she was very comfortable traveling to remote locations. However, what we found was that there was so little good forest that could still be saved. In the end, I went to twelve promising sites—in Sumatra, Java, Central Kalimantan, East Kalimantan, Sulawesi, Lombok, East Timor, West Papua, and Bali. I was looking for a place with a need for health care, an intact forest, and a supportive (and relatively functional) local government. The first one was easy: health care needs were essentially ubiquitous. But much of the forest was going or gone, even in protected areas, and local government capacity was often minimal. What we did find everywhere we went was the intense interweaving of human and environmental health. In some places, mercury used in small-scale gold mining was poisoning people and fish; in others, logging led to cholera outbreaks; and in extremely remote rainforest communities in East Kalimantan, we even saw the opposite: very healthy people in healthy forest.

I began to wonder if we would ever find the right place. One night in deep despair, I had a dream that gave me some hope. I was descending a spiral, eventually coming into a kind of chamber at the base. I had to reach my fingers through an intricate latticework mesh and try to crack a complex code. I knew that when I got it right, the mechanism would start to turn, and a golden model of a building would emerge from the smooth floor.

Finally, after almost exactly a year, I decided to visit the place that felt like my soul home—Gunung Palung National Park, where Cam and I had met. I had not gone there earlier because of the stories I'd heard of rampant illegal logging. The park had even been completely closed to researchers or visitors shortly after we left, because of so much logging. I figured that it, too, must be past saving. This would be a particularly horrible loss, since Gunung Palung is considered the jewel in the crown of Indonesia's national parks, with its diversity and beauty. But the previous month, Cam had accompanied a team of Indonesian scientists on the very first trip into the forest to assess the damage. From the very top of the mountain, he had cracked out his usually only-for-emergencies satellite phone and called me. With tears of joy, he told me that the core area around the research station was still intact. I immediately planned a trip to go see for myself.

On most of my trips, I was accompanied by Cam or Dr. Irene, but this time, I came with a new friend, Antonia (Toni) Gorog—a tall, blond woman who always seemed to be raising some baby animal she found. She had done doctoral research in ecology at Gunung Palung, along with her husband, Gary Paoli. Gary had served as manager of the Cabang Panti research station before I worked there, and their souls, like mine, had been captured by the place. I had heard so much about them, I felt as if I knew them for years before we actually met. Now our home base was just a few blocks from their house, not far from the botanical institute in the lovely town of Bogor in West Java.

The first stop Toni and I made was at the national park office in Ketapang (the nearest city to Gunung Palung), where we met the head of the park, Pak Anto. After hearing my idea of combining health care and conservation work, he was so supportive that he actually lent us a jeep and asked a staff member to assist us in exploring the area. So off we drove for the two-hour trip to the town of Sukadana—a town that lies right on the edge of Gunung Palung.

We had heard that Sukadana would soon become the capital of a new *kabupaten,* or regency. This was a critical piece of information, because most of the political power in Indonesia lies at the regency level.

The country is divided into ever-smaller administrative units, like Russian matryoshka dolls: the *provinsi* (province), the *kabupaten* ("regency," or county), the *kecamatan* (district), the *desa* (village), the *dusun* (hamlet), and the local neighborhood. Each of these levels has an official head, which simplifies passing information up (or usually down) the chain. Formerly, almost all the power was concentrated in the national capital, Jakarta; but starting in 1999, a process of rapid decentralization shifted much of the decision-making all the way down to the regency level. Basically, the process was something like states' rights gone wild. The head of a *kabupaten*, a *bupati*, is effectively a mini-sultan. That meant that the creation of a new *kabupaten* would give us the opportunity to work together with the recently formed local government from the very beginning. That would likely have its pros and cons, but overall, it seemed like a good opportunity.

While exploring the southern side of the park, we stopped at a roadside stand with a funnel and old cooking oil containers filled with gasoline—otherwise known as a gas station. This place also sold fried peanuts and iced tea in plastic bags with a straw stuck in, so we decided to take a break. Toni and I started talking to the woman who owned the stand. We told her how we had both done research many years ago in the forest, and this launched her into a long story about how much she appreciated the forest.

"We know if the forest gets all cut down, then less water flows down from the hills, and then we tend to get sick more often. You see the river over there?" We looked over and could just see the river at the bottom of steep banks with stilted houses along the edge. In front of each house was a small outhouse built on floating logs. She continued, "We drink from that river, and bathe in it, and if there isn't enough water, it's a real problem because people get sick. Plus, if the forest gets logged, we have more flooding." She then went on to tell that when her son had typhoid, they had been forced to cut down six ironwood trees to get him care. "God be praised, at least we were able to save his life."

Everywhere we went, Toni and I heard the same stories—much like

I remembered from 1993. The options for getting a large amount of cash quickly were very limited except by doing illegal logging within the national park. And health care was one of the few things you were willing to destroy your future to get. That woman's story also illustrated the negative cycle where everything was getting worse—more logging meant less water, which meant more likelihood of typhoid, which meant more logging. She fully understood this; she and her family just didn't have another choice.

Antonia and I then took a glorious hike up one of the mountains in Gunung Palung National Park. It was filled with life; we heard gibbons calling, found claw marks made by sun bears on the trees, and watched flocks of hornbills glide overhead. But there were also lots of stumps, and many of the biggest trees had been taken out. Reaching the top, we perched on a granite boulder and looked down over Sukadana. The sprawling town, nestled against the base of the mountain, stretches out to the South China Sea in two wings. Two large bays bracket the town, with a peninsula in the middle. We could see clearly how geography had helped make it an important sultanate in the 1800s.

As we sat on our rock, Toni told me excitedly about an idea she had come up with while listening to all the folks talk about wanting to protect the forest but also needing access to health care. She wondered if maybe logging and health care could be even more explicitly linked and villages could get discounts on care if the logging stopped.

I jumped in. "Great! This could engage social pressure to encourage people to stop logging. As long as we make sure that everyone can always get access to care through noncash payment options, this could work well. Otherwise, I'm not sure it would be ethical."

The idea of paying for care without money was one I had discovered on the trip I made across the United States shortly before moving to Indonesia. I was looking for a solution that would be more financially sustainable, but an experience after the tsunami had also strongly influenced me. I had been working in a tent providing medical care, where I examined an older patient with pneumonia. I prescribed him a very good, and very expensive, medicine. I was dumbfounded to see him walk out of the tent and throw the medicine away! When I ran after him to ask him why, his response was stunning: "Well, it's *free,* so it's obviously *no good.*"

No one likes to be considered poor, and people often suspect that they are getting worse care if they are paying less for it. In places like Serukam hospital that did sliding scales, doctors might also—without realizing it—give more attention to patients based on their payment status. Everyone, everywhere in the world, wants to be valued and treated with honor and respect. Giving free or discounted care to a singled-out group can make them feel devalued. It can also create a sense of entitlement. But I was at a loss for what system might work.

The answer came when I tracked down David Werner, author of *Where There Is No Doctor.* That book had proved extremely useful in the rainforest, not only in healing Tadyn's sliced thumb but also in diagnosing and curing my own malaria. When I visited him in his rickety cabin on a lake in New Hampshire, he gave me the best advice I had yet heard about how to both have a sustainable health care system and still make it affordable. But he did not invent the system, a village in Mexico did.

David had worked with a group of communities in an arid, mountainous region of the Sierra Madre, reached mainly by mule. He gave each a medical kit and a copy of his book in Spanish, but they were all tasked with finding a way to keep their kit stocked so the care would be sustainable. Each village tried a different system to pay for the replacement supplies. Some used a basic health insurance system, where people subscribed to get health care. This system had the same problem with health insurance systems anywhere: healthy people subsidize the care of the sick. This method tends to make well people a little resentful. Other villages used a pay-as-you-go system, but then some families were devastated financially.

One clever village invented a new system that I thought might work here, too, and that Toni and I had been talking about. In that village, if you wanted to pay with cash, you could pay with cash, but you could also choose to *work* to pay off the medicine. There were two work opportunities—either clinic work (cleaning, preparing materials such as rolling bandages), or caring for the fish pond (the fish were sold to pay for medicines). If the patient was too sick to do these things, a member of the family or even a friend could work in their place. This system was so effective and so well received by the community that all the other villages ended up converting to it.

Like me, Toni immediately saw the benefits of this system. The barter system allowed patients to maintain their dignity and get the care they needed. In addition, the whole community benefited, as the patients themselves helped to keep the health care system functioning. We agreed that we would have to see what the communities thought of these ideas.

That night, we visited Pak Mourni, one of the field assistants who used to work at Cabang Panti. Toni and I spread out a pile of donated eyeglasses on a floor mat, and everyone in the neighborhood showed up to see if they could find a pair that would work for them. It was such a joy when the perfect match occurred, after lots of laughing and trying on glasses. One gray-haired man, delighted with his "new" bottle-glass lenses, was mesmerized by the world around him. He couldn't stop staring at his arm: "I have *hair* on my arm! I had no idea. I can see every hair!" One woman was less thrilled. "My hands are so old! When did that happen?" A young man called out, "I can see to *Australia* with these glasses!" Giving the gift of vision is a bit like watching people fall in love—in love "at first sight."

Later that evening, about ten of us sat in a circle on the floor around the modest flicker of a kerosene wick. Everyone sang and made music, using spoons, plastic tubs, and a guitar. I leaned back and looked out through the open window to see hundreds of thousands of giant fruit bats flying toward the flowering durian trees in the hills near the sea. Watching wave after wave of their silent silhouettes against the deep blue-black sky, I was overwhelmed with a sense of peace and content-ment. I had felt greeted by these same animals the first time I came to Gunung Palung, and now I knew I was back home.

Some of the other sites I had seen in the last eleven months might have worked out, but they suffered from transportation difficulties or corruption or political instability that might have hamstrung the program. The villages around Gunung Palung seemed like an appropriate fit for this work; there were about 120,000 people without access to health care, and about half of them lived on the edge of an amazing forest that could still be saved. What's more, the national park office and the local government were eager for us to come.

Toni and I spent a hot, mosquito-filled night sleeping on Pak Mourni's thin kapok mattress, which he insisted we sleep on—while he and his wife slept on the wooden floor. There were no hotels in the area, but we could not continue staying with friends if they were going to give up their mattress. So the next day, Toni suggested we look at a house she had once visited, where a Dutch ornithologist lived with his field assistant's family. She thought we might be able to rent it from Pak Manto and that he could prove a good ally, with his impressive forest knowledge.

But when we arrived, we were sad to discover from his widow, Ibu Rajilah, that Pak Manto had died about a year previously. She showed me the precious X-ray they had traveled the two hours to Ketapang to obtain. It showed a hugely dilated heart that took up about three-quarters of his chest and fluid in his lungs. She also showed me the medicines he had been given: antibiotics, asthma medicine, Tylenol, and vitamins. Nothing that would have been of any use to him. Apparently, even if you had the money, you still couldn't get good health care.

We then asked about their children, and she told us her story, holding a little girl snuggled in her lap.

"My first three babies died in childbirth, and I almost died myself. After that, I adopted Eka, whose parents had ten children and couldn't afford to feed any more kids. Of course, I was afraid to get pregnant again, but it happened anyway. In those days, there was no birth control. Luckily, my husband had a job at the time, and we were able to deliver my baby with an operation in the hospital in Ketapang. My oldest was born—he is now fifteen and is married and has a one-year-old baby. A few years after that, I got pregnant again, but we didn't have any money, and again, the baby died when it was being born. Then my husband got a job as a research assistant, helping find woodpeckers in the forest, and I was able to have another operation. Our son Tri was born. Around that time, it became possible to get birth control, and I was so excited I got it the first week. But then my husband died. I still wanted to have a little girl, to have company in my old age. A woman I know died in childbirth—it's amazing that didn't happen to me. Her three kids were left without a mother, so I took her youngest girl, Puji, and now she lives with me." She indicated the girl in her lap by gently patting her.

Toni and I looked at each other. *This* is what it meant to have no access to medical care or birth control. Ibu Rajilah had lost four children in childbirth, delivered two safely with Caesarean sections, and adopted two—one of whom had lost her own mother in childbirth. And she lost her husband to a possibly preventable, or treatable, heart condition.

I rented Ibu Rajilah's house for one year, agreeing to her asking price without bargaining—two hundred dollars for the *year*—even though it was slightly above the going rate. She told us she was happy to rent it, as she usually slept at a little place she had built on the beach, to sell noodles and drinks to people enjoying the view. Even if we only used the house during our evaluation trips, it would be worth it, and it would help her out. I knew from my Aceh experience to be very careful about causing inflation, but this was close enough to the going rate, and I just couldn't bargain with her.

It was not only money she needed, though. She and the whole community needed access to high-quality health care—ideally, in a way that would also protect their long-term well-being. I hoped we could help them with that.

BY JANUARY, I WAS BACK in Sukadana, newly returned from a fundraising trip to the United States and happy to be moving into the rented house. Cam, however, was less happy. In fact, furious would be closer to the truth. He hated that I had chosen Gunung Palung. How could I go work at the place that it would rip his soul to pieces to watch be destroyed in front of his eyes? In fact, now he decided he didn't want me to do the work at all. This made no sense to me since it was the whole reason we had moved to Indonesia, but he was happy living on the island of Java and getting to work with so many Indonesian scientists. Why did I now have to drag him to Borneo, where everything would be more difficult?

"But, Cam, you have resisted every move we have made in our life together. You were so mad when I went to medical school in New Haven, you raged when we moved to California, and even though you yourself agreed to move to Indonesia, when it came to the actual leaving, you fell apart. Yet each time, you ended up loving the place and being so grateful for the new stage of our life. In the end, you have always thanked me. Maybe this is just the same?"

"No, this is different," he insisted. "I've never had to be in a place I love so much and had to listen to the horrible chain saws cutting it down every day!"

I saw his point. Still, I knew I had to try. In the end, all we could reach was a stalemate. I would start in Kalimantan, and he would stay in the city of Bogor. I hoped eventually he would come around as he had always done in the past, but I wasn't certain he would.

So off I went on my own—or rather not quite on my own. I took our helper Ratna with me for the first few weeks. The second day there, we spent the entire morning scrubbing the floor together. But when she went to dump a pail outside, I heard Ratna give a wild shriek. I hurried to her side, in my jeans with wet knees.

A dozen silvered leaf monkeys were jumping onto our thatch roof and scampering across it, eager to feast in the nearby rambutan trees. It was the first time in her life that Ratna had seen a troop of monkeys. She clung to me, scared but laughing. I laughed, too, and assured her they wouldn't hurt her.

I was still getting used to having helpers. From an Indonesian perspective, not having domestic help would be unforgivably selfish and miserly; if you have money, you must share it with as many people as you can afford to—and together, more can be accomplished. This was definitely the case, given that going grocery shopping required getting to the open market by 6:00 a.m., laundry had to be done by hand, hung on the line and quickly pulled in before the daily rains, water was carried up from the well, having chicken usually meant slaughtering it yourself, and even spices for cooking were hand-pounded. Ratna would soon go back to her family in Java, but she had found a neighbor, Yani, who would partner with me in this work. Already, I knew that my helpers should get joint credit for anything I managed to pull off, given that nothing besides housework would be possible without them.

Dr. Irene had been helping me start Health In Harmony's sister nonprofit in Indonesia. Part of the reason for this was legal, since foreign nonprofits are not allowed to conduct work in Indonesia, but the main reason was that I wanted local decision-making for any on-the-ground activities. Toni's driver had recently come up with the Indonesian name, and everyone I tried it out on loved it. The long version, Alam Sehat Lestari, could be translated as "healthy nature everlasting" or "nature health forever." But the real genius of the name came in the shortening, which tied them all together. In Indonesian, a shortening can take either the first letters or the last one or two letters, and they prefer if it also makes a word. Our short name would be ASRI (with the RI coming from the end of Lestari). ASRI, which is pronounced "ah-sree," means "harmoniously balanced."

Enough donors had chosen to support this work during my last trip back to the States to do the baseline survey and hire staff (barely), although only because I was working for free, and Cam was paying our living expenses from his Harvard salary. Luckily, I had scientist and doctor friends who understood the importance of getting good data before we started so that we could know our impact—if any. A beloved friend from Yale medical school, Alison Norris—with both an M.D. and a Ph.D. in epidemiology—had helped me write the baseline survey. The questionnaire covered a lot of ground: health topics, attitudes about the forest, and questions about household finances and livelihoods. The plan was to interview about 1,300 households in twenty-five villages over five weeks. Toni and I would spend a week training sixty-six nursing students who would conduct the survey as part of their required six-week community rotation. Half were coming from the Serukam nursing school (the hospital where I first thought about becoming a doctor), and the other half were from a nursing school in the town of Ketapang.

The logistics of the survey were already proving to be, without exaggeration, monumental. We had to get the teams of nurses to *all* the villages around the park—arranging buses or boats where possible, renting motorcycles for muddy roads, obtaining the list of households from each village chief for the random sample and, almost impossibly, finding lodging for every interviewer in someone's home. I had been working intensely on this over the previous few weeks, and in another few weeks, we would begin.

After lunch with Ratna, I got ready to go out to meet a man who, I hoped, might become an ally. Pak Bakhtiar was the second-in-command in the regency administration, and he was famed as someone who truly cared about both people and protecting the rainforest. He was also the head of the local chapter of the largest Muslim association in Indonesia.

I climbed into my rented car by twisting open the wire that held the door closed and then re-twisted it by reaching through the open window once I was in. I could literally see through the floor to the road below, and the brakes and steering left much to be desired. I was grateful I hadn't killed myself yet—or anyone else. While most things were extremely cheap, cars were expensive because there were so few of them. In Sukadana, a town of about twelve thousand people, there were only three other vehicles besides my rental, and all of them were falling-apart pickup trucks. Amazingly, one of my professors at Contra Costa Regional Medical Center, Kim Duir, had offered to match any donation to help buy us a car. I planned to name it after her.

I bumped along the dirt track, avoiding chickens, stray dogs, and the occasional cow, while a few bicycles zipped past me, deftly avoiding the car-size holes in the road. (Bicycles were definitely going to be a high-priority purchase.) Soon, I was settled on the wide wooden porch of a modest but comfortable house, with a geometrically carved railing that might have adorned a sultan's palace. In traditional style, the porch had no chairs. Pak Bakhtiar, a compact man in crisp white Islamic dress, was sitting cross-legged in the middle of his porch, surrounded by about ten local leaders who lounged against the wall of the house, some smoking their clove-scented cigarettes, talking and sipping glasses of reddish-colored sweet tea. It was a hot, lazy afternoon; cicadas buzzed in the many fruit trees around the house. I sat across from them against the carved front railing, facing these sunbaked, older men.

Behind me was the view of the same mountain that Toni and I had climbed together—but now with a big gash in the treescape, where someone had recently carved out a slash-and-burn rice field. Pointing to the bare area, Pak Bakhtiar spoke about the importance of protecting the national park and lamented how hard it was to enforce the restrictions with so few government resources. The men behind him, sitting against the dark wooden wall, nodded agreement, and one commented how all their water came from that hill and it needed to stay forested or the creeks would dry up. Pak Bakhtiar then invited me to share my own history with the area, what I planned to do, and why I was in Sukadana.

I began to talk about the time I had spent as a researcher at Gunung Palung and how it inspired me to start a clinic and work with the local communities to help save rainforests. I talked about how critical these forests were not just for local people but also for the health of the whole planet. One of the men leaned forward, in his handsome green batik shirt, and assured me that they knew that these forests were indeed the "lungs of the earth."

I agreed with him, and I was getting so carried away that I talked for quite a while, gesturing for emphasis and pointing to the forest that lay behind me. As I spoke, a brilliant blue butterfly landed on my out-stretched hand. Sometimes, butterflies like to lick the salt from your skin, but this one was just sitting on my hand, and I saw no reason to bother it. So I just held that hand still as I continued to talk, and the butterfly settled in. But after a while, I could see the men begin to murmur and shift uncomfortably. The last of the clove cigarettes were put out, and the glasses of tea sat untouched. Eventually, everyone just sat upright, staring at me. I fell silent, uncertain what was going on. The butterfly placidly waved its wings, reflecting blue iridescence with each wave, but it didn't fly off.

Finally, Pak Bakhtiar cleared his throat and tilted his head with its Islamic cap. He exchanged a look with the other men on the porch that I could not understand. He turned to me. "Do you know what we believe about people that butterflies choose to land on?"

"No," I replied nervously.

He paused and seemed to be considering. "We believe it is the mark of a holy person." After a deep breath, he concluded, "We will help you in your work."

Feeling flustered and not at all holy, I took a deep breath and silently thanked the universe. At that, the beautiful creature fluttered away.

9

Launching the Clinic

SUKADANA, WEST KALIMANTAN:
FEBRUARY 2006–SEPTEMBER 2007

"Hi, Irene. What's the news?"

I was sitting outside on the three rough-hewn logs that formed our back porch, enjoying the lovely view of the mountain across the valley. This call was only possible thanks to the cell phone towers installed in the past year. Without any way to communicate, starting a program in this area would have been unmanageable.

Though Dr. Irene would join the board of ASRI, I hadn't been able to convince her to come start the work with me in Sukadana because she had just gotten married, and she and her husband were both starting their residency training in Jakarta. She was helping find folks to work with me, though, and I suspected that's what she wanted to talk about.

"Kinari, I just talked to Hotlin, and she said you haven't called her yet. I told you, you *have* to call her. She is the perfect person for your program."

"But, Irene, I just don't think a *dentist* is one of the first people I should hire. I need a medical doctor and nurses first. I just wish you could come."

"I know, Kinari, but I can't join you right now. You have to listen to me. I want you to hang up right now and *call her.* You will never find someone better for helping you run this program than Hotlin. She's just back from studying in England, and now she isn't sure what she's going to do with her life. She used to run a clinic *on a boat,* in Sumatra, so she's got good experience. *Call her!* She's perfect."

"Okay, okay, I'll call her right now."

I called Hotlin. We talked for more than an hour. The roosters were crowing so loudly it was sometimes hard to hear, and I kept walking around the house to get a better signal.

I told her about my vision for this work. I explained how, as doctors, we were taught to see people as just assembled body parts—in her case, the teeth—and not holistically. I told her about people having to log to pay for medical care and how the health of the environment affected human health. I believed we should not only be treating people as whole human beings but also expand the circle out and treat the community and the environment. I didn't agree with just putting Band-Aids on people's medical problems; I thought we should be treating the causes of the medical problems as well.

Hotlin agreed with me completely, and she talked about how much she wanted to do community development as well as medicine. She was less certain about the environmental side, but she was willing to accept it. We discovered that we had both gone to Aceh, in different locations, to help after the tsunami. She had experienced the same frustration at seeing the big nonprofits not listening to the communities.

The two of us were getting along so well on the phone, and I was so impressed with her experience that, before ending the call, I asked her if she would like to help me turn this vision into reality. Hotlin paused, and then she accepted the job, right there on the phone. She told me she could come in March. I told her that would work perfectly, because I should be done with the baseline survey by then. And then I had to swallow hard.

"Uh, Hotlin, there's just one thing. I have very little money from donations." I took a deep breath and then named a salary that was one-third to one-fifth of what dentists were making at that time.

Silence and a long pause as she considered. Then she said, "Okay, I'll do it. I will need to get my family's approval, but I'm pretty sure they will give it to me."

I hung up and danced a little jig. Irene was right: Hotlin was going to be perfect.

IN FEBRUARY 2006, DR. HOTLIN arrived along with Dr. Romi, a medical doctor whom Dr. Irene had found. Both proved to be absolutely lovely human beings—perfect first employees. Toni and I had finished the base-line survey, and we found that 79 percent of the people thought the park was a good thing and 83 percent wanted it to be there for future generations. They were desperately poor, though, and we estimated from the population size that there were at least 1,350 households logging full-time with many more likely logging part-time and not admitting to that on the survey.

With this data and all the conversations we were having, it was clear that health care was a key need for stopping the logging. But we knew we would need more formal listening sessions with the communities to determine what else they needed.

Cam had still not decided to join me, but we talked often by phone, and I hoped that he would come around. In the meantime, we were both extremely busy as he worked on various projects with Indonesian scientists and Romi, Hotlin, and I prepared to open a clinic. Our first difficulty had been finding a building, one that was big enough that we could convert into a clinic *and* that was occupied by people who would be willing to move out. When I had just about given up in despair, we finally managed to persuade the owners of a nice-size house, likely because they had two children with medical needs. Hotlin, Romi, and I then spent our days painting, sanding, and overseeing the renovation, when we weren't driving the two hours to Ketapang to fill out yet more paperwork for the permit or to buy supplies and equipment.

In June, after nearly four months of struggling with logistics, we decided to take a trip to a village to distribute eyeglasses and listen to people. The village we visited that day was surrounded on three sides by big hills that rose from the plain like green dollops of mashed potatoes. Even though these forest-covered hills lay within the protected national park, the forest was patchy, and as we drove up, we could see the bare ground of cleared areas where people were growing bananas on the steep slopes in areas that had been completely logged. Shortly after we arrived, many people gathered, and Hotlin and I helped people try glasses. Romi diagnosed one person with leprosy, two with chronic obstructive pulmonary disease (their lungs destroyed by smoking and cooking on open wood fires), and a few children with malnutrition. Since at that time we didn't yet have a clinic permit, all we could do was refer them to the government health center, even though everyone was complaining about the poor quality of care there. We also asked the crowd what they would think of a noncash payment system at our clinic. The response, like in the other villages where we had tested the idea, was vociferously positive.

Packing up, we had been showered with the joyful thanks of people who now had glasses and were able to see clearly for the first time in years (or maybe for the first time ever). We shook hands with everyone, touching hand to heart each time to convey our trust and affection. But just as we were about to leave, a woman in a torn T-shirt and faded sarong rushed up to us, sweat dripping off her face. "Thank goodness I got here in time. Would you *please* come and see my son? He is very sick."

How could we say no? We all bundled into our recently purchased used Toyota called Kadie (KD), in honor of Kim Duir, my professor in residency who had, indeed, successfully solicited and matched donations for it. I invited the anxious mother to sit next to me in the back seat, showing her how to close the door and put on her seat belt, as she had never been in a car before. The woman's house was about a ten-minute drive. She must have run the whole way to have covered that distance in the time since we had arrived in the village. The tiny house was built on stilts over a bit of cleared mangrove swamp, out of old worn wood and thatch. We walked to the house on a boardwalk of single two-by-fours. The main room was barely six feet across. There was no furniture—only a few of the handwoven reed mats called *tikars*. But the room was spotlessly clean, and there was something about the way the bits of multicolored salvaged wood had been put together to make the walls that told me that this woman had an eye for beauty.

A smaller room had been created in a back corner of the main room. The mother gestured for us to enter it, holding back the dark blanket that hung in the doorway. Inside this completely dark space, we found a seventeen-year-old boy, squatting on the floor and rocking back and forth in agony. He begged us to put the curtain down, turning his face away from the sliver of light that entered the room.

His mother told us their story. Dardi had spent the previous *eight years* in that room. He could not stand any light or noise; he was in constant pain from a pounding headache. Only through the greatest effort could his mother convince him to eat something each day, and he would still vomit. They had been to see many doctors that had cost all the money she had, but none of them had been able to figure out what was wrong with him. With questioning, we learned that he was also incontinent. This surprised us given the lack of smell—in fact, he seemed meticulously clean. But then we noticed that he was squatting over a hole in the floor that had not at first been visible in the dark.

Romi and I did not have our stethoscopes or ophthalmoscopes, so we arranged to come back the following day. On the way home, we discussed the possibilities.

"It could be a brain tumor, but in that case, it's amazing he hasn't died by now."

"Maybe it's a bacterial abscess, but again, he would have died. And he should have a fever, which he doesn't."

"What about neurocysticercosis, from eating uncooked pork? Although the presentation doesn't look like it—and he is a Muslim, so that makes it less likely. Maybe some other parasite in his brain?"

"Do you think it could be meningeal tuberculosis? The slow course would make sense, although I would expect him to be skinnier—but maybe his mom is doing a really good job feeding him. I can't believe we forgot to ask if he has a cough!"

We discussed if there was any way we could avoid a CT scan, since getting to the nearest one in Pontianak would take a car trip, a five-hour boat ride, and then another car. Given his flinching if we just touched him or spoke loudly, we couldn't imagine he could tolerate the three-day trip. And then, of course, there was the cost.

As soon as we got home, I emailed one of my professors from residency, neurosurgeon Laszlo Tamas. It took about an hour to send a short email over my cell phone, but we badly needed a consult. The next day, I had an email back from Lazlo. "I've been studying the literature—and I don't mean the recent literature, because there is none—but the literature from the 1920s and '30s, which was the last time anyone in the U.S. wrote an article about tubercular meningitis." He shared a fascinating finding that could help with the diagnosis: patients with brain tumors often have trouble looking upward and downward, but with tuberculosis, they can have difficulty with looking left and right. Lazlo also said that his bet was on tuberculosis (maybe with a tubercular abscess), because the patient definitely should have died by now with a brain tumor.

Armed with this information, and with all the diagnostic equipment we had at the time (a reflex hammer, an ophthalmoscope for looking in the back of the eye, and a stethoscope), we showed up again a few days later at Dardi's house. Hotlin and Toni came along, too, out of sympathy. (Toni was a regular visitor from her home in Bogor, helping with conservation planning and coordinating grant writing.) We were all desperately hoping that it would turn out to be tuberculosis, because that could be relatively easily treated.

Dr. Romi, ever so gently and slowly, approached Dardi. He did an excellent job of getting a history and performing a thorough neurologic exam. In a way, it was a relief, after spending so much time on paperwork for the clinic, to be actually seeing a patient—but so sad to be seeing one like him. Dardi cowered in the corner, afraid of the light, afraid of the noise we would make, and clearly in horrible pain. It turned out that Dardi did have a cough—he had even been coughing up blood—although it took quite some probing to get the information, as he clearly thought it was irrelevant and a distraction from his real problem: his aching head. On physical exam, he had noises in his lungs indicating disease, the backs of his eyes were bulging out (an expected effect of increased pressure in his brain), and—surprise, surprise—he had trouble looking to each side, though he could look up and down with ease.

After our exam, we sat down with his mother and his married older sister, making sure that Dardi could hear us from behind his dark curtain. We told them we were pretty sure that he had tuberculosis of his brain—and that this might sound like bad news, but it was actually good news because it was treatable. He might need to take medications for an entire year, but we were hopeful that he could be completely cured if he took the medication faithfully and did not stop partway through the treatment.

Dardi agreed to try, although he seemed pretty skeptical, and he was very worried about the cost. We told him not to worry. We had already thought that one of the noncash payment options we would accept were handicrafts, so we told him that his mother could pay by making some of her lovely woven mats. The look of happiness that passed over Dardi's mother's face was beautiful to see. She wouldn't have to accept charity—but she would be able to pay for her son's care and still feed the family.

Romi and I discussed it and decided that we should just go ahead and treat him even though we didn't yet have all the official paperwork for the clinic. The problem was that with tuberculosis of the brain, sometimes the patient needed intravenous steroids for the first few days of treatment because all the dying bacteria can cause more brain swelling as the immune system responds. Steroids would calm down the immune system. While we wouldn't be able to do the IV steroids in his home, it was possible he wouldn't need them. Dr. Romi was already treating one other patient, a young schoolteacher, who had another unusual form of tuberculosis—this time, in her belly. She had been seen in the hospital in Pontianak, where they had missed the diagnosis. So much fluid had accumulated that she appeared to be seven months pregnant, and we had been afraid if she weren't treated soon, she would die. Within days of starting treatment, her belly started shrinking, and she now looked normal. It is possible that because of her we were feeling a little cocky. So Romi prepared the medicine and we brought it out to Dardi.

But the following day, we found out we had made a mistake. His mother pulled up to the clinic on a rented motorcycle to say that they had to stop the medications because he had started vomiting uncontrollably soon after starting them. It appeared treating him would not be so easy after all. His intense reaction was in one sense a good sign, though, because it meant we had the diagnosis right. But he was going to need steroids, as well as close monitoring, during the first week or so of treatment.

Clearly, he needed to be admitted as an inpatient—but we didn't yet have a functional clinic with a permit. We explained the situation to his mother, and soon we made another trip to Ketapang to try yet again to get permission to open.

This time, Dr. Romi went into the permit office by himself, and after just a few minutes, he came out fuming. One of the women in charge of writing the required letter was clearly so fed up that she finally told him plainly, "Look, you just don't get it, do you? Why do you think I keep making you redo almost every letter? You paid the permit fee at *that* desk over there, not at mine. If you had paid it at *my* desk, I'm sure I could write you this letter right away. But since you didn't, I may not get to it for a few months—in fact, I may *never* get to it!"

When Romi related this experience, Hotlin became furious, but I considered giving in. "Maybe we should just pay the bribe. Think about all the patients like Dardi who aren't getting care because we don't have a permit."

Shooting fire from her eyes, Hotlin pointed her finger at me. "Absolutely not! We will *never* pay a bribe! The minute we start paying them, we will never be able to stop. We have to find another way around this." I had been the one in the beginning arguing we should conduct all our work without bribes and Hotlin had been skeptical, but now she held me to task, and I was grateful.

The next day, Hotlin, Romi, and I went to visit Pak Bakhtiar on the same porch where that iridescent butterfly had landed on me. It was shortly before Friday prayers. We explained our dilemma and asked what he thought we should do. We were barely able to finish before he gave an exasperated snort, grabbed his phone, and ordered his secretary to get the number of the office where we were having trouble. "Go in on Monday," he told us. "You'll get your permit. These officials make me so mad! Don't they care about the people?"

On Monday, Hotlin went into the office this time. Just as she walked through the door, someone called out, "She's one of the doctors from ASRI. Quick, take her in to see the director!" Hotlin was ushered into the director's office. Sitting behind his large desk in his pristine batik uniform, he stumbled over himself, apologizing that the system for issuing permits was new, and he was so sorry that some of his staff didn't appear to know how to do their job yet. He himself would personally make sure we got our official paperwork as soon as possible.

A few weeks later, we were finally allowed to open our doors to patients. While the delay in getting the permit had been incredibly frustrating, in truth, we had actually needed most of that time to set up the building, set up the systems, and hire staff: three nurses, a pharmacy assistant, and a driver to go pick up medicines. We had also used the World Health Organization's list of essential pharmaceuticals to order all the medicines and created both a paper medical record and a digital one.

That last part was possible because I had finally convinced Cam to move to Borneo. At one point, I had called my mother in despair that it seemed like he would never choose to come on his own. She told me that I had to make him move and that if I didn't, I would lose my marriage. So I had girded my loins and prepared to fight. But our daily call that night was a bit of an anticlimax, and in some ways, it felt like he had just been waiting for me to insist. I flew to Java, and shortly before our twelfth wedding anniversary, I packed up his whole life and had it shipped to Borneo—he simply acquiesced and got on the plane with me. The first few months, with the daily sound of chain saws, were indeed a rough adjustment for him, but he also loved the hikes in the forest and magical sunset swims. I think for both of us, snuggling together at night was healing, and it felt good to be finally making this project happen together. I was so grateful when he was also willing to use his fabulous technical skills to create the electronic clinic database.

During the final weeks before we opened the clinic, we were also negotiating with Dardi. We needed him to come to the clinic and sleep there for at least a week so we could safely start his medicines, but he was terrified to leave the house. After much discussion, he finally agreed to a detailed transportation procedure: we would pick him up after dark, giving him dark sunglasses to wear (against the light from motorcycles); we would have a special pot for him to urinate in (he refused a catheter); and we would give him both painkillers and anti-anxiety medications.

On our opening day, July 15, 2007, our first two patients were the children of the wonderful couple who had rented us the clinic. Toni flew in, and we all celebrated together. Then on day two, after seeing a few patients during the day, we blackened the windows in a back room of the clinic and set out to fetch Dardi. When we arrived just after dark, he was sitting huddled in the front room of the house—possibly for the first time in eight years—holding a small bag of possessions and wearing a windbreaker, shaking with fear. Neighbors arrived to help. After waiting for the antianxiety and pain medication to kick in, they very carefully carried him out to the car over the rickety planks forming a bridge over the swamp. Once in the car, sitting between me and his mother, and, with his dark glasses settled in place, Dardi held my hand in a viselike grip.

Arriving at the clinic, he was half carried inside, although he was able to put some weight on his legs. I was surprised to find that he was a bit taller than I. As we tucked him into his room in the back, he seemed content and pleased to have made it. We decided to let him rest that night and start the medications in the morning.

Day one of his treatment, he was in less pain and had no vomiting; day two, he started being able to tolerate light; day three, he could get up to go to the bathroom with help. And day four, he walked out of the clinic on his own in the middle of the day and got into the car to be taken home.

Success, right? Not quite. Two months later, Dardi was complaining about the medications. There were too many. He was already better; he was urinating just a few times a day, he didn't have a headache, he liked to listen to the radio, he wasn't coughing anymore. Why should he still take these medicines?

He was not the only patient with these questions. In just the first two months of the clinic, we started about ten patients on tuberculosis medications, and half of them already wanted to stop their drugs. This is a problem all over the world; it is very hard to take four different drugs for six months or more when you feel cured after a month. But if you stop before the end of the course, the bacilli become resistant, and they can come back with a vengeance.

We scrambled to start a program called Directly Observed Therapy Shortcourse (DOTS), where people are hired to watch patients actually take their medicines three times a week. To start this, we needed to know approximately how many tuberculosis patients were in the region, to help us estimate how many DOTS workers we would need. We had made an appointment with the head of the government clinic and met with him in his office. He looked me up and down lewdly and complimented me on the modesty of my dress. He then said he was very sorry, but he couldn't tell us how many TB patients they had. He had political aspirations, and he had been reporting that he had eradicated tuberculosis from the region. He would appreciate it if we did not report that we had *any* tuberculosis patients. We gently told him that our case reports would always be accurate, nothing more and nothing less than what we found.

Despite not having an accurate count, we decided to hire eighteen women who lived in villages scattered around the park. But we were racing against time. Dardi had refused to take his medications for two weeks, despite three visits from us in that time. Two weeks is right on the line for the maximum time to skip treatment without developing resistance, and we kept our fingers crossed that the medicines would still work. Luckily, the woman who became Dardi's DOTS worker just happened to be young, charming, and extremely pretty; he was putty in her hands and never missed another dose.

Fascinatingly, one of the other women we hired as a DOTS worker was Ati, Jono's wife, who had accompanied Jennifer and me on that first trip into the forest. I had thought at that time that she would have made a great doctor—but that had been an impossibility for her. Still, by using the Indonesian version of *Where There Is No Doctor* that I had given her, she had become a critical health care resource for her community. Her copy was barely holding together, so she was delighted to get a new one when we gave the book to all the women we hired as community health workers. Now she would be an even more critical medical resource for her community.

Dardi and the community health workers were an excellent lesson for us in how solving one problem might just cause another. Leaving the dark and coming into the light never seems to be an easy process. However, we also found that as we addressed any given issue, it might also have positive ramifications beyond what we anticipated. It was already clear, though, that everything we did was going to require persistence.

10

Radical Listening

SUKADANA, WEST KALIMANTAN:
SEPTEMBER 2007–APRIL 2008

O ne of the key steps in starting the clinic had been to set up
the barter payment system. Word had spread like wildfire that
this was an option, but figuring out the details was tricky. We
had, after much "market analysis" and community feedback, set prices
for the handicrafts that we planned to sell in the United States and the
seedlings we would use for reforestation (grant funding for reforestation was much easier to come by than money for health care). We also
determined wages for working in the clinic, doing laundry, cleaning, or
rolling bandages and cotton balls. However, I was worried that these
options wouldn't be enough, and I wanted to also give people the option to grow vegetables. This would solve our problem of how hard it
was to get good-quality vegetables to feed the staff every day. For some
unfathomable reason to me, nearly all the vegetables were imported
from Java.

This plan sparked a big fight with Cam.

"You'll either be flooded by people wanting to work, or you won't
have enough people. Don't do it!"

We went back and forth for about twenty minutes as I tried to explain to him all the reasons I thought it would work. Before long, we started yelling, and I'm sure all the neighbors heard us.

"You don't get to tell me what to do! I never tell you how to do your research. You didn't even want to come, and now you think you can tell me how to do my work? You're not trying to figure out a solution, you're just dictating!" I shouted.

"Well, if you don't want my help at all, then why are you asking me to design the clinic database and help plan reforestation work? You can't have it both ways!" Cam fired back. "And you're the one who made me come here. I feel like I want to scream when I hear the chain saws, and what if they are now logging the trees we love most at the research station?"

"Well, we have to try to save it! We can't just give up. At least I'm trying! But I can't do it if you just oppose everything I attempt because you're sure it's going to fail."

Finally, we agreed to a truce where he let me make my own decisions about how the program would go—although begrudgingly—and I agreed that I would give him full control when I wanted his help. He felt this as a loss of his romantic idea that we would do the work together, but I felt a little space to breathe.

However, if I was going to start a garden, we would need land, which I also knew we would eventually need for our own clinic building. I found donors, and then, unexpectedly, we were contacted by an Indonesian soil expert, Bang Luth. He had found us online while finishing his Ph.D. in Paris, and he wondered whether we happened to need his skills. Even more amazingly, when he came, the piece of land he considered best for a garden was just two doors away from the house we had rented as our new clinic. And more remarkable still, the government official who owned it agreed to sell it to us for what we had raised, even though we found out later it was less than he paid for it.

Luckily for my marriage, the garden actually worked out very well. We hired a young man who had previously been a logger to oversee it, and nearly every day, he had folks coming to pay for their health care—not too many and not too few.

I HAD BEEN LISTENING TO the community's needs and their ideas about the root causes of logging since my very first trip to Indonesia—this is how I knew that they desperately needed health care. But now I planned to begin a very formal and structured listening process in every *desa* around the park. I wanted to know what the communities saw as the solutions to stopping the logging (or if they even thought that was possible). The baseline survey indicated that the majority of people cared about the forest, and another small survey we had done just of loggers found that 99 percent of them would prefer to do other work. But logging in and around the park was rampant, and chain saws could be heard every single day in every village, so I wasn't sure how receptive people would be to finding solutions to protect the forest. Still, I wanted to try.

When I explained to the team my desire to do this, everyone but Hotlin expressed some skepticism. "What if they ask for things we couldn't possibly provide? Would we be setting up false expectations that they could ask for anything? Shouldn't we instead get some experts to come in and do an assessment since village people are unlikely to be able to determine what the solutions are? And besides, it would be embarrassing to say that we didn't know what the solutions are—after all, we are more educated than most people around here."

I heard them and admitted it was a risk, but having Hotlin's support that the communities *were* in fact the experts, not us, strengthened my conviction that this was the right way to go. While we now had a number of local staff, most of our team had been brought in from outside our regency because medical training was extremely rare locally. For example, Ibu Irawati, our pharmacy assistant, was the *only* person with this training among the entire 120,000 people in the regency (and there were no pharmacists). I wanted to ask the people who were directly experiencing the problems what the solutions were.

Our next hire was someone to work full-time on the listening sessions. Farizal was the perfect person, having grown up in this area. Years before, he had been hired by Toni and her husband, Gary, to work as a field assistant at the Cabang Panti research station. Gary and Toni had been so impressed with him that they raised money from friends and family to help him complete high school and college. And fortuitously, we were able to hire him immediately after he finished his undergraduate degree in sociology. The only problem was that I couldn't make him stop working seven days a week and long into the night, since he was so thrilled to be able to use his knowledge to help his home region.

I assumed that every village might have very specific local solutions, so we would need a lot of meetings. Farizal set up these gatherings by first going to the head of the village and explaining how we wanted to talk with people from the community about their well-being and about logging. He would ask the leader to make sure that many women, heads of community groups such as farmer or fishing cooperatives, and religious leaders were invited.

We held the first meeting not too far from my house one evening after clinic, and some of the team joined. Farizal, Hotlin, Romi, Irawati, and two of our nurses, Wil and Ika, and I sat on woven mats on the cracked cement floor of a small hut as more and more people kept crowding in. This house had been selected for the meeting because it was the largest in this little community, and it was the only one with electricity. The smell of dust from the dirt road floated into the room, mingling with the scent of freshly bathed, hardworking bodies. It began to feel pleasantly warm against the slight chill of the tropical evening.

Hotlin brought the meeting to order and made the introductions. She then explained that we wanted to hear about people's hopes and aspirations and to talk about what they saw as the solutions to make their lives better and protect the environment. She emphasized that they were the experts, not us. We knew that they not only understood the reasons for the logging but also might know how to stop it.

Our head nurse, Pak Wil, then spoke from the heart about his own painful experience—growing up in another part of Borneo surrounded by forest, only to see it all cut down for palm oil plantations. He talked about the worsening health, the contamination of the water, and the sadness of the people when the beauty of the forest was gone. "Please believe me, it is possible to lose all your forest, and it is horrible when it happens."

He then shared his own internet research into understanding climate change and described how important rainforests are for keeping the climate stable all over the world. Wil had in fact become passionate about climate change and had been looking up information on weekends during shopping trips to town. He explained the importance of the carbon stored in trees and how Indonesia's deforestation was a driver of global climate change. He used an evocative metaphor: the polar caps are like two moons on the top and the bottom of the earth, which ought to be full moons of ice but were already waning to half-moons; if all the rainforest in Indonesia got cut down, they might go to *new moons*. If that happened, much of Indonesia would be underwater, and there would likely also be huge effects on the health of people all over the world. It was a poetic and powerful way to describe what was happening, and there were nods of understanding all around.

After this introduction, the team had decided that I should ask the key question. They thought that it would be good to have this question come from someone who clearly came from the wider world. Pak Wil passed it over to me, and after thanking everyone for gathering, I got to the heart of the issue. "We are here today to ask you a very important question. You all are guardians of a rainforest that, as you have heard, is important to the whole world. What would you all like to receive from the world community, as a sign of thanks, so that you could protect this forest? I can't promise you that we can definitely do everything that you need, but I can promise that we can try." We had played with the wording of the question beforehand. We wanted it to be as open as possible, to have it be framed in a positive light, even despite the rampant logging, and make it clear that the exchange would be mutual gift-giving (reciprocity), not a monetary transaction (as payment for ecosystem service is usually set up). We recognized that, like most humans on the planet, they didn't want to be destroying their home and future but simply didn't have a choice. Most people globally are in this same situation. But with the right knowledge and resources, most of us might be able to make different choices. In the framing of our question, we wanted to make sure it was clear we recognized them as equal partners: they had a gift to give the world as guardians of this precious rainforest, and they might receive gifts in return.

As this question settled in with the group of about fifty gathered in a circle and some even leaning in through the windows, people smiled, sat or stood up straighter with pride, looked at each other with nods, and then began to really think.

A slender middle-aged village leader, wearing a government-issued khaki shirt, spoke first. "We are very grateful you have asked us this question. We also really want to save the forest, but it is very hard to know how to get all the loggers to stop, even if they mostly want to. To tell you the truth, I used to be an illegal logger as well, but I realized that logging was really the same as drilling holes in the bottom of my own boat. The thing is, we are all in the same boat, so I understand why we should work together to keep it afloat."

Next to speak was an older man, somewhat bent over, dressed in clean but faded clothing. "I am very worried for our children. When I was little, all the kids could name at least fifty types of trees, but now the children only know a few. We have less rain, and the dry seasons are longer now. Then, sometimes we get bad floods, and that never used to happen. The weather is no longer balanced the way it used to be. When I was little, we also had much more water flowing down from the mountain, and we rarely got sick. Now we get sick often. I knew that our forest was important for our own health—but I didn't realize it could affect the health of people in other places far away. I am glad that maybe those people are willing to work with us to try to find ways to stop the logging. Thank you for bringing us this news. I am happy that you understand that medicine can sometimes cure disease, but only a healthy environment can give us health."

There were many nods of assents to this speech.

One of the farmers offered his opinion. "It's like this. You need to understand that there are two reasons the forest is now so much farther away from our village than it used to be. The first reason is that many of us have been doing illegal logging. But the other cause is the way we farm. Many of us do shifting cultivation. We burn down one area of forest and plant some rice and then let it grow back for about seven years before burning it again. The problem is that you can only do this a few times, and you get less rice each time—and it doesn't taste as good. This system worked well for our ancestors, because there was lots of forest and not many people, but it isn't working anymore. We have to find a new solution. We have heard that there is a way to plant in one place without buying expensive chemical fertilizers, but we don't know anyone who knows how to do this. I think we need to learn this system. Maybe this is something we could get help with. We also used to get many of our vegetables from the forest, but now that isn't possible, and few of us know how to grow them. The other reason many people are logging is that they have to travel so far to go to a hospital, and even then, they often don't get well, and they have to spend even more money to try somewhere else. I think we need a good quality hospital here."

A fragile young woman, cradling a three-year-old, spoke up. "That's true. We really need better health care here. Until you opened the clinic, we had very few options. In April, I had a miscarriage, and I almost died from the bleeding. There was no way I could have made it to Ketapang on the back of a motorcycle, I definitely would have fallen off and died when I was bleeding so much. Luckily, Dr. Kinari picked me up with her car and brought me to care, but we really need to have an ambulance that we can rely on. The government clinic has one, but it usually doesn't have gas or we have to find a driver ourselves."

Everyone was silent for a few minutes, mourning with her.

We continued to listen for more than an hour until it became clear that the key issues had been discussed and there was a sense of completion in the group. There is often a kind of magical moment when the energy drops in the room and it's clear from everyone's body language that everyone agrees with the solutions they have come to.

"So if we are hearing you correctly, it sounds like the critical thank-yous that you would need from the world community, to help you stop logging, would be access to high-quality health care that you can actually afford and training in organic farming so that you don't have to do shifting agriculture and so that the loggers can get money in another way."

There were general murmurs of consent, so I continued, "We would never want to deny health care to anyone, but we want to provide a special benefit when people stop logging. What about having the world community pay part of your treatment costs if you come from a non-logging village? Even if you come from a logging village, you can still pay for your care with noncash means. If you decided to pay with tree seedlings, for example, you would have to grow *more* of them if you came from a logging village—but you would still get your health care without spending any money. That way, everyone would always be able to access care."

There was strong agreement. One man said he loved the idea because "this way, I might finally be able to get my neighbor to stop logging!"

People were very excited, and they were still discussing the ideas as they left the meeting. I walked home by moonlight, digesting everything we had learned. Once home, I found Cam holed up in his office trying to finish a National Science Foundation grant application. He would probably be working late into the night, and he might have to drive the two hours into Ketapang in the morning to submit it, given how bad the internet connection was over our mobile phones. He was right: being a world-class scientist in a remote village in Borneo was not proving easy.

While he worked, I sat in the dark with a cup of tea, looking across the valley at a ridge in the park framed by stars and thinking about what we had learned that day and what it meant for Gunung Palung. People had been so clear that they were often forced to destroy their own futures to get their very basic needs met. On a broader scale, I knew this was largely because resources had been taken from these communities through a long history of colonization. But could I actually help bring what was needed back to these communities? Obviously, I wouldn't be doing it on my own, but my team had been right; it was stressful to ask for what they needed, trust they were right, and then have the faith that we would be able to help them enact those solutions. The next question, too, was whether every community around the park would come up with different solutions. If we had to deal with many different issues in different areas, it would be even harder. I knew there was only one way to find out—we would have to ask.

BY MARCH, WE HAD CONDUCTED these listening meetings with every *desa* around the park, and they were going surprisingly well. I somewhat jokingly started calling the process of listening *radical*. First, because we were doing it with radical love and respect for the communities. And then, the really radical part was that actually having the communities design the program subverted the normal power structure between outsider projects and local communities. Finally, I liked this word that comes from the Latin word for *root*: listening to the root of the problem and roots of the solutions.

My worries about every community coming to different solutions turned out to be entirely unfounded. There had been, in fact, complete consensus. But now the details needed to be worked out. To do this, we had two meetings that gathered groups of village leaders from each half of the park. In the area of health care, the consensus was that ambulance service and mobile clinic visits were needed, since travel imposed a huge cost, and many people had no access to transportation at all. We had to have some negotiation at that point. For example, the initial request was for weekly mobile clinic visits to every village around the park, but we explained that it was impossible to have enough staff for that. However, we told them that we could probably manage twice-weekly trips—eight every month—to a rotating list of centrally located spots they could choose. They were very happy with this solution. They also ideally wanted a hospital but were content with expanding the clinic services that we were already providing when we said we could not build a hospital anytime soon. They would be delighted, too, to get organic farming training as soon as possible. They were happy to bring together people from many villages to learn together.

There was one additional request that we heard a number of times: please teach the children about conservation so that the impacts would be long-lasting. However, since there was another nonprofit that occasionally worked in the schools, the leaders agreed that we could leave it to them for now.

In the details meeting with village leaders, we also discussed how exactly the incentives for protecting the forest would work. We found that we were always the ones offering *more* incentives, while the participants would argue that the rules should be stricter. It was generally agreed that for health care access, residents of a village that were not logging should get a 70 percent discount on their bill. We then proposed that villages that were still logging should also get a discount, 30 percent, since they had agreed to work with us to try to stop the logging. Villages that had not agreed to work with us, as well as patients coming from areas far from the national park, would pay full price.

We were very clear, however: everyone could always come to the clinic, and all the surrounding villages could call for the ambulance, and anyone could pay with noncash means—no matter where they came from. The communities argued that only the villages that were not doing any logging should get mobile clinic visits and sustainable agriculture training, but in the end, we decided that all the villages would get those things, because the logging villages might need them the most.

However, I was very worried about how we were possibly going to get the funding for all their requests. The truth was that, at this stage, the "world community" we talked about was really just me, my friends, my father's friends, and a few other generous donors. I also knew that I couldn't possibly do enough fundraising in the United States and help run the program here. So far, Julia Riseman had been volunteering in the role of executive director for Health In Harmony, but we needed a paid person. Without someone to help us get the word out more widely, we would never be able to expand our circle of support. An unexpected solution presented itself.

For one of the radical listening meetings, in the town of Siduk, a volunteer, Brita Johnson, joined us. Hotlin and I had alternated translating for her as she took copious notes. Afterward, Brita had bubbled with excitement. "I've heard so many nonprofits talk about 'community organizing' and 'listening to communities,' but they so rarely actually do it. You *do* it! Once you asked them, they were able to figure out exactly what the solutions should be—and you could tell how excited they were, that it might really be possible to implement those ideas. If the world partners with them, these issues might be solved!"

We were both struck by the fact that the problem wasn't actually money in their pockets. Even if people had lots of cash, the things they needed weren't available locally.

"Kinari, I think in America, people would jump on this idea of being able to save rainforests by providing the health care and organic farming training that these communities have asked for. It's brilliant. I think you're going to get lots of support! We just have to figure out how to get the message out."

Then I had a thought. "Brita, you wouldn't by any chance be interested in working for Health In Harmony and helping us seek those thank-yous from the world community, would you?"

Brita laughed and said that, if we advertised the position, she would apply in ten seconds flat. While I wanted to hire Brita, or someone like her, as a full-time executive director for Health In Harmony, I knew we had only a few months of pay in the bank. We would just have to hope that the new person could raise the money for their own salary plus what we needed to meet the community requests.

Given that the national park held truly huge amounts of carbon, and protecting it would indeed help the whole world, it might be possible at some point to get funding through carbon financing. This would mean being paid for trees we planted, since they would absorb large amounts of carbon over their lifetime, and potentially to receive funds for protecting standing forests. Despite covering less acreage, Indonesian forests in totality actually have more carbon in them than the Amazon. This is because of the high density of large trees in these forests and because of deep peat deposits in the ground (the densely packed leaves in the ground are an early stage of petroleum). Stopping deforestation globally would get us 20–30 percent of the way to a stable climate, so there was a huge incentive for the world to help stop the logging. But there wasn't yet a clear and easy carbon market for this kind of work.

Ideally, the funds would come from pollution emitters. This could be done on a national level, on a company level, or even personally. In this way, the amount of carbon emitted, or carbon footprint, could be "offset" by paying someone else to stop, or avoid, emitting carbon. Given that current levels of pollution are unsustainable, this could just be a dangerous self-deception and keep polluters from changing their behavior, but if it would help the communities around Gunung Palung, I suspected I would be willing to swallow my qualms about the system. Besides, it was hard to be too moralistic about it since use of fossil fuels is simply unavoidable with our current technology—and I had done more than my fair share of international flights. However, to get something like this set up would take a long time, and the whole world had not yet settled on a good system. For now, we were just going to have to rely on donors.

Brita and I began stacking the plastic chairs and plastic boxes we had used to bring the banana-leaf-wrapped snacks into our new four-wheel-drive truck. As we did this, I had a moment of disbelief looking at our new truck and all the equipment bought with our first grant from the U.S. Fish and Wildlife Service: *Did we really manage to make this happen?* Maybe it would actually be possible to meet the community's needs.

My worries were further calmed by thinking about Sheryl Osborne, who had become Health In Harmony's biggest single donor. My favorite neurosurgeon professor from Contra Costa had introduced us, guessing that we might really like each other, since her charitable foundation focused on both great ape conservation *and* maternal and child health. Sure enough, we adored each other. Sheryl could see the carbon benefits, but what really motivated her was the incredible biodiversity of the forest and especially the more than three thousand orangutans in and around the park. She even joined me on my third scouting trip to Gunung Palung. It had helped the work immensely to have a confident, well-dressed, white-haired woman stand next to me, conveying to government officials that I had backing—even if she and I both knew it was a bit of a bluff.

Maybe there were other people around the world who also cared, if they were just informed. We would have to see.

11

Returning "Home"

SPEEDBOAT FROM PONTIANAK TO TELUK MELANO
AND CABANG PANTI RESEARCH STATION: MAY 2008

A re we all set to go into the forest next week?" I had to shout over the noise of the speedboat engine. Cam and I were sitting in the two front seats next to the boatman, who perched up on the edge of the speedboat, steering the boat with his toes while he smoked yet another clove-scented *kretek* cigarette. The boat was hydroplaning so high that its bow hid our view of the mangroves skimming by.

I was back from a month-long visit to the United States, where I had been surprised by how receptive people were to partnering with these communities and helping with a new experiment in conservation and health care. This isn't to say we were swimming in money, but thanks to Brita's help, combined with grants, we had raised enough to increase salaries of the Indonesian staff up to standard wages and hopefully soon start the organic farming training and the discounts for health care based on logging. The path forward was starting to feel a little more navigable. The final step would be to sign the reciprocity agreements. We had a big meeting planned for the next month to do this, but it seemed like it would just be a formality since all the other meetings had gone so well.

"Yep, all set," Cam responded with a squeeze of my hand. We had met up in Pontianak, as he was also returning from a research trip on a different Indonesian island for his new National Science Foundation grant. "It's been so long since you were up there. Apparently, there is almost nothing left of the buildings, but I hear that one of them still has enough of a floor for us to camp on if we put a tarp up."

We were planning to go back to our forest research station, Cabang Panti, the following week. I hadn't been back in twelve years. I'd lived in Sukadana for more than a year now—but I hadn't been able to obtain a permit to go into the forest itself. It was frustrating to be so close but unable to visit the place that felt most like home to me. Odd that a rainforest on the opposite side of the planet from where I had been born would feel like home, but it did. It was my soul home.

Speedboats skimmed past us in the opposite direction, toward Pontianak. I also saw one of the old *Express* wooden boats we used to take, put-putting along, but its top deck was mostly full of motorcycles now. All the fast boats were crammed with at least thirty passengers, and our own fiberglass roof was so laden with luggage that every wave caused it to bow in alarmingly.

After crossing a bit of the open ocean, the boat slowed to enter the estuary of the giant river that comes down from Gunung Palung past Teluk Melano. The quieter engine made it easier to hear, and it also brought the driver's cigarette smoke curling around us. The enormous steel bridge was still there, but now it had roads connecting it on both sides—though they still didn't extend as far as Pontianak. The strangest thing to me was the transformation of the town itself; it was filled with huge rectangular cement structures that looked like small New York tenements rising from the edge of the water. Weirdly, these buildings didn't house people but instead were fake caves for millions of swiftlets, whose nests are the prized, gelatinous ingredient in bird's nest soup—and demand was soaring in China.

Speaking over the engine, I asked Cam, "Can you believe that Melano now has over twenty-three thousand people? It must have had about three thousand when we first came."

"Why are people so unwilling to talk about population growth?" he replied. "Dad told me that when he was growing up on Mount Kilimanjaro, there were four million people in Kenya, but there are now forty million. On his last trip, he said the road that used to be miles of Serengeti was now miles of slums."

"It *is* amazing that our parents are likely to be the only generation there will ever be who will experience *two* doublings in their lifetimes—if the world gets to eight billion people."

"Did I tell you, Kinari, on this last trip, I calculated my own doubling year? When I'm forty-five, the world population will be twice what it was when I was born." Referring to a climate ecologist friend of his, Jason Bradford, who was part of the Post Carbon Institute, Cam noted, "But Jason says there isn't a single climate model that says the world can hold eight billion—so who knows what's going to happen. We are probably living at peak, and the crash will happen soon."

I countered, "But we can't forget that the problem isn't just people, it's consumption. One American uses as many resources as three hundred seventy Tanzanians. Americans having fewer kids, or using fewer resources, is probably more important for the sustainability of our planet. Although I recognize population matters, too."

"Well, if you keep saving those babies, our earth is going to be in even worse shape. *Every* human further destroys nature. And I know what you are going to say—don't say it!"

Cam knew that I would remind him of the data that the only way to reduce population growth is to give people the faith that their kids will survive, but he had a hard time believing it would balance out. Whenever we argued about this, I remembered an experience I had had at Serukam hospital during my last year of medical school that had convinced me that the data was correct. At the time, I was helping measure the impact of their mobile clinic trips. I found that the mortality rate of children under age five in most of the villages before they started their work was 25 percent. Right around then, a local woman I knew well asked me how many children I wanted to have, and I had answered, "Two."

Her stunned response had been: "What are you talking about? You can't have *two*! What if one of them dies? What if they both die? You have to have at least four to be sure that two will survive. It's true, two would be ideal, but that is just too dangerous!"

Given the historical mortality rate, she was right; out of four children, one would very likely die, and there was a good chance that two would—but on average, three children out of the four would survive. She was overestimating a bit, to hedge her bets. But I had also shown that the infant mortality rate had been dropping in that area, down to 12 percent, and it probably continued to drop even more over the years as it had been doing globally. I had recently heard from that same woman, and I was surprised she had just two children—the number she wanted. Even if they are not consciously aware of it, people probably have a very good intuitive sense of how likely it is for a child to die, based on how many of their friends have lost children. Then, as long as they have access to birth control, they can have the number of children they want.

Cam, following exactly where my thinking was going after so many years together, added in appeasement, "At least you are distributing free birth control. Without that, no one can have fewer children."

The whole conversation made me uncomfortable, though, given that we were trying to get pregnant ourselves. Would he love a child if we had one? I knew he had a deep love and concern for both people and the planet—but, in the nature-human divide, there was no question which side my husband was on. I sometimes thought he was truly happy only when he was in the forest. And I got it. I felt that if we lost the rainforest, I might not be able to go on.

In Borneo, there was no avoiding deforestation. Nearly every day, it was in our face. We loved a gorgeous fig tree, home to a giant flying squirrel that would glide out at dusk, swooping over the forest—and then, one day, that tree and all its neighbors were gone. It was an experience that recurred constantly as we watched the fastest rate of deforestation that world had ever known.

Cam looked at me. "I can barely face going up to the research station. Do you think it will all be logged?"

The last reliable information we had was from the trip two years earlier, when Cam had trekked into Gunung Palung and found it still pristine. As a sign of his support of our work, the head of the national park, Pak Anto, had given me permission to accompany Cam on his new research visa, which would allow him to reassess his plots. Cam wanted to see if he would be able to get another round of data from the tree plots he'd put in for his Ph.D. Two of the staff from the park were also planning to come with us, to evaluate whether it would be possible to open up the area for research again.

Our boat pulled up to the dock in Melano, where a crowd of dozens of people were gathered to meet the passengers. I heard people calling to me, "Hallo, Doc! Welcome home!" Some of the greetings were coming from people I only vaguely recognized but who clearly knew me, either from the clinic or the radical listening meetings.

After bumping to a stop and tying up the boat, we clambered over the windshield, finding a precarious foothold on its bow. Pak Agus, ASRI's driver, was there to meet us, and he reached down from the dock, which, at low tide, was high above us, to help me climb up. An unwelcome new sight was the trash floating on the river, visible below the stilts of the boardwalk. On my first trip, fifteen years ago, only leaves or paper were used for packaging. Now there were black plastic bags, candy wrappers, and cigarette packages everywhere. People hadn't yet figured out that plastic didn't just degrade like banana leaves. I wondered how long it would have taken me to independently deduce that fact if I hadn't been told. And even when they did figure it out, what would they do without a landfill? It reminded me that while billions had been lifted out of extreme poverty over the prior decades, much of that change happened through the use of fossil fuels—and at the expense of the natural world.

We walked through the market, a riot of color and smells, with fruit stands, cheap clothes hanging on racks, stores selling myriad varieties of pungent salted fish, and goats wandering in between everything while munching on plastic. We saw uncountable swiftlets swooping over the river, around the bridge, and over our heads, on the way to and from those tenement birdhouses that had been constructed for them. The screech of the birds, combined with electronic birdcalls and shouts from the market, overpowered conversation.

Many of the concrete birdhouse blocks appeared to have windows, but on closer inspection, one could see that they were just painted on. Selling those incredibly expensive nests might seem to be a sustainable way to make money. But these ten-story cement buildings required huge amounts of wood for their frames—and these days, that wood was mostly coming out of the national park. Hotlin had talked with a Chinese-Indonesian shop owner about building the birdhouses with steel, and she learned that the cost would actually be about the same or even less—but the problem was that no one there knew how to build with steel.

Still in his black mood, Cam commented that Chinese medicine was probably the major cause for loss of biodiversity worldwide and a disaster waiting to happen in terms of transmission of diseases from wild animals to humans. "At least the invention of Viagra stopped some of it. But how anyone can believe that eating those slimy nests will prolong their life, when there is *zero* data to support it, is beyond me."

Maybe he was right, but I also knew that people weren't always aware of the consequences of their behavior. People in California living in a concrete home probably have no idea that the plywood for the forms came from the rainforests of Borneo. The guests at a Chinese wedding might not know where their soup comes from or what it took to obtain it. And how many people at a doughnut shop would stop to think that the palm oil used for frying may cause the effective extinction of orangutans by 2023 through fragmentation and forest loss?

Before setting off, Pak Agus pulled out his cell phone to coordinate with the clinic, to pick up some seedlings grown by a patient as payment for medical care. Sixteen years ago, we could never have imagined that Melano would change so much. And having 3G capacity in the cell phone towers was even more unexpected. (Of course, the internet barely existed in 1993—let alone cell phones!) Cam marveled at how easy our phone made these simple tasks. "Maybe technology *will* help," he mused. "There might be a way for people to actually know the implications of their choices and even partner around the world to change our trajectory. Who knows? Maybe I'm wrong, and we'll be happily surprised."

THE NEXT WEEK, CAM AND I were dropped off at the point where a new bridge crossed a river flowing down from Gunung Palung to meet the national park staff who would accompany us to Cabang Panti. This was the same river route I had first traveled so painfully fifteen years before, but now we could start farther along, thanks to the road. I had made the trip quite a few more times after 1993, though not in the past twelve years. Cam looked at his GPS. "Can you believe it's only fourteen miles as the crow flies? Back then, it felt like the edge of the earth."

The national park office boat pulled up under the bridge, and we climbed into it. The brand-new fiberglass boat was much faster than the old wooden ones and much easier to maneuver over the sandbanks. In this part of the river, the bordering nipa palms looked the same as ever—but wherever a dock had been built, I could see through the gap that there was almost no forest behind. A troop of proboscis monkeys traveled along the remaining strip of trees beside the river. The males have preposterously huge noses, suggesting that the females believe that old wives' tale about the size of the nose relating to the size of the penis. Those peculiar-looking monkeys always made me laugh, and I was especially happy to see that they were still here in the mangrove swamps.

But as we traveled farther upriver, the mangroves petered out. We eventually reached a stretch of the narrowed river that I seemed to recognize—but yet, it couldn't be.

"Cam, *this* can't be where the forest used to start to arch over the river, can it?" Cam consulted his GPS and nodded dejectedly. I asked them to stop the engine for a minute, and as the boat slowed, I climbed on the bow to get a better view. On either side of the river, all I could see was grassland—the awful, invasive alang-alang sword grass that looks so green and lush but catches fire every year and is almost impossible to eradicate. This wasteland vegetation can't even be eaten by animals. How many thousands of species had once inhabited these acres? Now the diversity was reduced to maybe ten species: the grass, a few worms, some insects, maybe two or three birds. Nothing else could live there.

Looking up from his GPS, Cam spoke as if half in the past. "This is the area that I wrote you about in that letter. Do you remember the time I came down from the forest when you were in your last year of college? This whole area was burning, and it felt like paddling through Dante's *Inferno*. Burning trees were falling, we could barely breathe from the smoke, and the monkeys were screaming. I'll never forget that hell."

Cam looked up at me with weary eyes. I got back down into the boat so we could continue upstream toward the mountain and, possibly, the intact forest. We held hands.

Farther upriver, we at last could see the forest closing in over our heads. And although some of the big trees had been taken out, the forest was basically intact. Rainforest that hasn't been too heavily logged can recover—if it doesn't burn. But if it burns too much, it will eventually turn into that awful grassland—which is exactly why our reforestation efforts were focusing on those areas.

At the end of the previous year, with Cam's advice, we had started planting trees in one of those grassland areas in the national park that had been logged soon after I left in 1994. The first time I surveyed the site, I stood on a giant charred stump—fifteen feet across, with its buttresses—and all I could see around me were grass, ferns, and the occasional woody snag. When the reforestation team prepared the site, we were distressed to find not a single natural-growing seedling beneath the grass and ferns. Even though the area bordered directly on a still-intact area of forest, no seeds had germinated in that heavily logged area. This may have been because the seed dispersers (birds and mammals) just wouldn't go into these areas, and seeds do not last long in the tropics. The few seedlings that might have managed to sprout were likely killed when the area burned every year. Planting those trees had been amazingly healing for me—and for the whole community. Even if it would take hundreds or even thousands of years for our planted area to look the same as uncut forest, we were already restoring habitat; each year, there were more bird species in the reforested area, and we were even seeing telltale signs of mammals, such as giant bearded pigs and monkeys.

Reforestation was important for the communities as well. The former village head told me how angry they all were when outsiders came in to log the national park. Finally, he said, they decided to "drink the red bowl." The Dayak men tie a band of red cloth around their heads, drink a blood-colored potion from a red bowl, and then literally run *amok* (a useful Indonesian word we have borrowed into English). I have heard different versions of this story, but some people say that the logging company finally left when a head was taken. Given that the traditional practice of headhunting resurged in Borneo in 1998, during a widespread conflict over land rights, that account is possible. Regardless, the story showed the community's deep opposition to logging.

As we continued, Cam and I were relieved to see that the forest looked progressively less damaged. The problem was that any logging at all would have destroyed Cam's research into the evolution and maintenance of such amazing diversity. Long-term rainforest studies are rare, precisely because of the likelihood of losing research sites to fire or logging.

The new fiberglass boat was impressively fast, and soon we passed the fork where two branches meet: the Air Merah (Red Water), and the Air Putih (Clear Water). The Air Merah flows through more peat swamps than the Air Putih, and its red coloring comes from tannins that stain the water—much like making a cup of sun tea. Farther downriver, the water can be so darkly stained that it is nearly black by the time it flows into the ocean. Even the South China Sea is golden-colored for many miles out, due to the many millions of cups of tannin "tea" that are poured into it daily.

I felt my heart begin to race as we progressed farther upstream. Could anyone else ever understand the intense love, and worry, that Cam and I felt for this place? When our boat finally neared the area of the research station, Cam and I stared upriver, almost holding our breaths. The sight of the smooth red-barked *Tristania* trees, curving down over the river banks, brought back to me all the emotions of living in the forest—the painful journey through my own soul, as well as the intense feelings of love, forgiveness, and belonging I had experienced there. I think of that year as the *beginning*. It has shaped every moment of my life since: the beginning of my soul's awakening, the beginning of a life path that I never imagined before that year, and the beginning of learning to listen. In the deep quiet of the forest, the silence spoke, and I started to pay attention. In being still, I began a journey of listening to memories, to nature, to the Divine, to others, and finally to my own needs and desires. That ranking was pretty much the increasing order of difficulty for me, and I knew I still had quite a bit of work on the last one.

Finally, there was the sign announcing that we were entering the Cabang Panti research area. And below it was a new sign: *Awas: pohon ini dipaku.* "Beware: these trees have nails." It must have been posted by the last researchers to leave. Most of the trees had tree tags with a nail, and maybe that is all it meant, but it could have scared loggers whose chain saws could dangerously kick back if they met metal. The difference was dramatic; beyond that sign, there had been no logging at all. *None*. It would have been easy enough to bring wood out along this river, but there they stood—enormous trees, right next to the river. Maybe the signs had worked or maybe the loggers knew that, in some strange way, this area was sacred to the international community. Cam and I began to breathe. Then we began to rejoice. We couldn't wait to run around and greet all the trees that were our old friends.

The boat cruised past the ruined remains of the field assistants' bunkhouse. A few rusted pieces of old tin were propped over a new sapling structure, and there was a fire pit suggesting hunters had been camping there. Now I was flooded with a new fear—that, even if the trees were still there, all the wildlife would be gone. The boat curved upward, gunning through the rapids and around the rocks. At one point, Cam reached over to push us off a rock, and a few minutes later, we had to get out to coax the boat up over a partially submerged log.

With a few more bends and turns, we arrived at the closest of the group of research station buildings. During my first year, this house was unoccupied, but I knew it well. The ironwood posts were still there, though most of the rest of the wood in the building was gone. The two national park staffers got to work salvaging what they could, trying to create some kind of usable shelter. Cam and I started walking up the trail to check on the rest of the buildings. As we walked along, I found myself putting my hands on beloved trees in greeting. But something was strange: it all looked the same. Not a little bit the same but exactly the same. The light in certain places was just as I remembered it, the trees were absurdly familiar, and even the undergrowth seemed completely unchanged. I thought I must be remembering incorrectly.

Arriving at the main camp area, we were saddened by the devastation. With no one there for more than ten years, the forest (and possibly hunters) had destroyed the entire building. Twenty years of research, now trash: the shells of old Macintosh computers (the very first desktop model); rusting batteries; a few remaining bits of roof tin; blue plastic barrels that had contained collated specimens, now cracked open. It was hard to tell where the various parts of the building had been, until Cam pointed out an odd-shaped tree that we both remembered. That tree had grown up from a fruit seed tossed outside by some researcher years before, taking root under the wooden ladder we used for making repairs on the roof. The sapling had grown quickly, likely because it was feasting on the extra light around the camp. Potentially with the help of a bored researcher, the sapling had grown zigzagging between two rungs, making an *S* shape. The ladder and the building were gone, but the *S*-shaped tree was still there. Would this bend in the tree and some scraps of plastic soon be all that was left of the research station?

We continued upstream toward the three tiny huts that Cam, Alex, and I had called home in 1993 and 1994. Of Alex's hut, there was nothing left at all, and Cam's was gone, too, except for the ironwood posts. He turned off to examine the site, while I walked along the trail toward my own hut. I imagined that he would also be remembering our first kiss there that had turned feverish in its intensity. My heart had raced with an incredible sense of nervous anticipation—like a door to a new world had been opening. And indeed, that is what had happened. His love remained one of the key pillars of my life.

Just before the small trail to my hut, I passed one of my favorite trees. The enormous strangler fig wrapping around the huge trunk looked the same as before. The fig had started from a tiny seed high up in the canopy, and it sent its tendrils down to form an intricate lattice that by now enveloped the tree—and might eventually kill it. But what was odd, once again, was that the tree and the fig looked *exactly* the same. They didn't seem to have grown at all. Walking down to where my hut had been, I found that, like the others, everything was gone but the posts.

As I stood where the platform had been, I realized I was looking at the little tree that used to grow right through it. A notch had been cut for it through the floor and roof, and here it still was, having outlasted the building. But in this case, I could be certain: the tree was exactly the same size as twelve years ago. My hands still easily stretched around it, and there was the twig where a little tree shrew used to jump at dawn, on its morning rounds. That shrew had never once spared a glance at me, just going busily on its way. Another branch, where I used to hang my flashlight, was still exactly there, and it, too, had grown no bigger. That was impossible. Twelve years—and no change at all?

Cam walked up and read my thoughts. "It's eerie, isn't it? It feels like we never left, or we just stepped away for a moment. Like a huge chunk of our life went by, but the forest stood still. I recognized a seedling along the trail, and it's still a seedling, exactly the same size. Our huts have crumbled to dust within an apparently unchanged forest. I should have expected this, because the measurements in my plots showed that the trees were only growing a millimeter a year in diameter, and the seedlings only grew a centimeter a year in height. But somehow, being here, this really drives it home."

I looked down at a seedling at my foot—one I had almost stepped on unaware. "You're telling me that this seedling could be forty years old?"

I reached out and touched the little tree that used to go through my hut, wrapping my fingers around its trunk. "How old does that make *this* tree? Hundreds of years old? What about that giant tree over there with the strangler fig? How many thousands of years old do you think it is?"

Cam lapsed into scientist mode. "It's hard to say. It's likely that growth happens in huge leaps, when they get extra light. You can see how fast the trees have been growing in the reforestation site. The problem is, we just don't have long-term data—and you can't read tree rings on tropical trees, because they grow all year round, so it's hard to know what's going on."

"Or *not* grow all year round, as the case may be." I smiled at him.

I remembered the day an orangutan I had been following for days had traveled right over my hut without pausing or seeming to even notice it. That was when it really sunk in that our camp was just like everywhere else in the forest: we barely made a blip on the terrain. And now, the forest had almost completely dissolved our existence without seeming to stir itself at all. My whole life could fly by, and this little tree would barely be any bigger.

If—if it were left alone.

Even though we humans seemed so insignificant in this other world, where time passed in a completely different dimension, I knew that this was an illusion. In three chops of an ax, or two seconds with a chain saw, this tree's life would be gone. A tree several meters in diameter that might be many, many thousands of years old could be cut down in a few minutes. How long would it take for forest like this to recover, if it were logged? Could it ever really recover?

This disconcerting feeling reminded me of fossil hunting with my father as a teenager. I found a bone from about twenty million years ago, and later, we identified the bone in my dad's collection whose shape was the most similar: a dog bone. My father held up the dog bone in the anatomically correct position, while indicating the dog's height with his other hand. I held up the fossil bone next to it—and then in unison, our heads tilted back and our mouths fell open as we stared up beyond the ceiling. Our imaginations were filling in the enormous size of this prehistoric creature. What had been just a bone a moment before had become a fearsome giant (it was a *Hemicyon*: a huge ancestor of both the dog and the bear). Now looking at that sapling and guessing how old it was made me feel once again that same warping of time and space before my eyes.

Cam and I walked farther upriver. After just a few minutes, I heard a branch snap and a tree rustle—the movement of an orangutan. Each of the larger arboreal species moves in a different way, and during my time in the forest, I had learned to tell them apart by the sound. Peering through the trees, we could see the young male moving away from us.

If these trees were thousands of years old, how many generations of orangutans had been feeding on the very same trees? No wonder they could find them so easily! Maybe each new generation added only a few more trees to their repertoire.

A short while later, off to our left, we heard a snuffling grunt, followed by something crashing through the underbrush. Recognizing the unmistakable sounds of the giant Bornean wild boar, we smiled at each other and walked on. Later, as we approached the big bend in the river, Cam reached out a hand to stop me and silently pointed ahead. A whole family of otters was playing in a small tributary—three babies were rolling and frolicking in the water, while their parents rooted in the banks for food, putting their long whiskers to good use. None of them seemed to notice us as we watched their sleek bodies glide through the water and playfully twist and turn around each other. In the distance, we heard the high-pitched chattering of red leaf monkeys as they leaped from tree to tree. Cam and I were overjoyed to see and hear so much wildlife in such a short time. Not only were the trees still here, there seemed to be even more wildlife than we remembered.

As we rounded the bend in the trail, we came upon one of my favorite places on the planet—the place where I had immersed myself, after seeing the universe in a totally different way. In this spot, the river glides over smooth granite boulders before making a bend and tumbling into a wide pool of crystal blue with a sandy bottom. On the far side of the river stand about ten emergent trees, their enormous crowns arching above the shorter trees below. Many of these huge giants have trunks six to ten feet across, with great buttresses extending even farther. Their tops could be 250 feet high, 21 stories. The tallest trees in the world, the coastal redwoods of the Pacific Northwest, stand 330 feet, 30 stories; but, as conifers, they somehow seemed not as impressive, with their conical shape that stuck just one terminal point up high. These rainforest beauties are shaped like giant oaks, sending out massive branches impossibly high up. Seeing them again reminded me of the brilliant line by Alice Walker: "Ecstasy is uncut forest and the smell of fresh baked bread."

We stood, nearly overcome in that ecstasy. Then Cam turned and swung me into his arms while we both laughed out loud. For the last few years, we had been acting on faith that whatever was left of the forest was still worth saving, even if partially logged. The relief was profound, to discover that this was still some of the best forest left anywhere in the world. All those years that the researchers spent at camp, they worried that they were doing nothing for conservation. But simply by being there, they may have been saving the last intact watershed of lowland rainforest in all of Indonesian Borneo.

12

Deaths and Resurrections

SUKADANA, WEST KALIMANTAN:

JUNE–JULY 2008

otlin paled as the man continued to shout at her. At least one hundred people—including nineteen out of the twenty-one village leaders from around the park—had gathered for the community meeting where we had planned to sign documents with each of the leaders that spelled out the details of what each party would agree to. This meeting, which we thought was going to just be a formality after all the calm and collaborative meetings we had already held, was not turning out that way.

We were gathering just up the street from the clinic under an incongruous yellow tent with blue frills that we had rented from a wedding vendor. One of the hills of the national park was visible even from under the tent, and as usual, the whine of chain saws could be heard in the forest.

"You can't ask us to sign these agreements!" the headman from Penjalaan village stood to yell. "The district head has to sign first! You have to follow procedure!" He took his seat in his green plastic chair, fuming.

Hotlin responded calmly, "There is no need for shouting. We have spoken with the district head, and he agrees that the village leaders should sign first." She paused and smiled. "Besides, we don't believe in top-down work. We believe in bottom-up solutions. If you all don't agree with the proposal, then it doesn't matter what anyone else thinks. *You* are the important players here." She turned to address the whole group. "If you don't want our help, that is the way it is. If you cut down all your forest until you have no water to drink, and you all get sick, that's your decision. I will go home to Sumatra. But if you want us to help you implement your solutions, we are ready to do that."

Then three representatives of another village stood up and shouted that they would be very angry with their own village leader if he *didn't* sign. "If you sign, we will get extra discounts in the clinic! We will get the organic farming training we asked for. And all you have to do is say you will *try to* stop the logging. You don't even have to actually *do* it!"

The headman from Penjalaan stood up again and started shouting back at them: village leaders had to follow protocol, and who were they to tell him what to do?

Hotlin, Farizal, and I looked at each other, wide-eyed. What was going on? Why were we getting so much anger *now,* after a whole year of productive meetings with communities around the park? The village head from Penjalaan was enraged way out of proportion to his complaints. Anger is rarely expressed overtly in Indonesia, so this was even more upsetting. It was obvious that he was not giving his real reasons; it was clearly not a question of procedure, as he claimed. After yelling at Hotlin, he had angrily turned away and lit up a cigarette, in blatant disregard of our request that no one smoke in our meetings. Everyone was clearly upset.

A very old chief then stood up and tried to calm people down. "Now, now, there is no reason for fighting here. ASRI simply wants to help us. Every village that signs the agreement will still get a 30 percent discount off the cost of health care, even if there is still logging happening in that village, and non-logging villages will get a 70 percent discount. Why wouldn't you want your villages to get these discounts? Let's just take a little break now, so everyone can take a deep breath."

But the break was not exactly a period of calm, as the factions jostled for position. Hotlin, Farizal, and I stood to one side, feeling stunned. Then I saw one young man leave his group, as they gathered around the table that held the coffee and snacks wrapped in banana leaves, to come over to us. In hushed tones, he explained that the village leader from Penjalaan was, in fact, one of the logging bosses. He explained that the man was probably afraid that if he *did* sign the agreement, his village would be angry at him if he continued to log in the national park. And if he *didn't* sign, they would be angry for not having access to the extra incentives. His only hope was to convince all the village leaders to refuse to sign on some procedural technicality. Evidently, we had put the man in a no-win dilemma.

After thanking the young man for explaining the situation, we called the meeting back to order. Only about half the people sat down again; the rest stood around the edges of the tent, just inside its shade. Again, there was yelling on all sides. We knew what the problem was now, but we still didn't know how to fix it. Fortunately, someone else did. The old man who had called the break was sitting near the front, and he stood up wearily and said, "Farizal, why don't you come see each one of us individually, and we can sign or not sign, depending on our own conscience. I, for one, will be at my house tonight, and I am happy to sign."

We nodded in agreement. This solution was a clever one, even if it would take more time and cost more money. Hotlin closed the meeting with the announcement that Farizal would visit each one of the village heads. Yet no one got up to leave, and there were rustlings and mutterings all about. Finally, one young man came up to me and tucked a note into my hand. It said simply, "We want to sign."

I went up to Hotlin, slumped at the table at the front of the tent, and showed her the note. She and Farizal consulted together, and then she picked up the microphone to announce that some villages wanted to sign, and anyone who wanted to could come up and do so now.

The village leader from Penjalaan immediately let out a furious bellow. "How dare you say that people can sign! You already closed the meeting. *No one can sign now!*"

I stepped to Hotlin's side. "Okay," I said, "no harm done. We will do this slowly. And remember, this is up to you; if you don't want our help, that is fine. We will just leave."

But lying in bed later that night, it didn't feel fine. Would the opposition win out? Would the delay mean that those opposed would have time to lobby other leaders not to sign? Had we approached this in the wrong way, even to ask for signatures? A single logging leader would be able to hold his whole village hostage—and might even be able to stop us completely. I knew that we were supposed to honor the *tuan rumah* (literally, the master of the household)—that is, give respect to the official head of any organization, area, or institution. To show appropriate respect, we had decided to ask for the headmen's signatures. But what if they all refused to sign? What if only one or two headmen signed?

The hardest part was that I really believed in radical listening. If the communities didn't want the program, then that was probably the right conclusion, but this felt like a few people, with their own interests at stake, hijacking the situation. If the majority of people wanted the program, would their voices win out?

Tucked into my mosquito net, I began to cry and found I couldn't stop. Cam was away on one of his frequent trips to give scientific talks. Now I wondered if he was right and the whole thing was a ridiculous idea that would never work. Had I made a horrible mistake to give up a comfortable, rich life in the West to try to do something this difficult? I was tired, bone tired, and if we had to give up at this point, I would feel like a failure, like I had somehow betrayed all the people in the villages and our donors who believed we could do this. Sometimes it felt like I was trying to wrest this program out of the mud, grabbing and pulling and shaping it with sheer willpower. Some part of me knew that it had nothing to do with my own strength and that I was not actually in charge; but it also seemed that if I relaxed my attention for even a moment, it would all just melt back into the earth.

I tried to live by Mahatma Gandhi's principle—that only the means, and not the ends, were in my hands. But that night, I had to face the fact that, despite doing the best "means" I knew how, what I felt like I was being asked to do just might not work out. Recently, a visiting friend (who was also my professor) had mocked my sense of a calling. Faith, he insisted, was not necessary to do this kind of work. My answer was simple. "Well, *you* might be able to do something like this without believing in something greater than yourself, but *I* can't. Sometimes it just seems too impossible, too hard, too painful, too unlikely, and the only thing that gets me through are the affirmations that this is the path that I am somehow 'supposed' to be on." Now, however, I wondered if I was deluding myself and maybe, I, like Cam, would end up losing my faith. After we married, he had moved away from the evangelical Protestant religion of his childhood, and we had become members of a Quaker and later Mennonite congregation. But by medical school, he rarely attended services and, of late, his anger at spirituality in any form had been increasing. He had even taken to calling religion "psychological child abuse." I agreed with him that the idea of a dominating male God who would punish people for eternity was a terrible one and should not be perpetuated, yet I still usually believed in that sense of all-suffusing love and interconnection that I had experienced in the forest.

But that night, I felt alone in the universe, forsaken. My sense of despair might have been exacerbated because in addition to feeling like I was failing this program, having a child was also not working out. We were continuing to try, and now after three years, we still had not gotten pregnant. Being so frequently apart was likely part of the problem, but still, I felt it should have happened by now. Had I given up my one chance to have a child?

I cried myself to sleep and then woke up in the middle of the night with what Cam calls the "threebeejeebies"—that 3:00 a.m. sense of dread, self-doubt, and anxiety. In those dark nights of the soul, it is hard to remember that there are ever butterflies.

THE NEXT MORNING, I WOKE to the sound of gibbons singing behind our house. Every morning we could hear them in the hills around the village, but it was a special treat to hear them so close. In the light of the new day, I wondered whether the intense opposition we had experienced in yesterday's meeting might be a good sign because it suggested people were worried the system would actually force them to stop logging. I still felt lost and alone and would have loved to stay in bed—but when you work at a clinic, you can't not show up. Patients were already waiting, and maybe there would be the satisfaction of helping heal someone or at least comforting a patient if we couldn't.

Before 8:00 a.m., I was bicycling up the dirt road to the clinic. Our fifteen or so staff members had already gathered for the morning meeting. A slew of bicycles and motorcycles, belonging to patients and staff, were lined up under our thatched roof "garage." I nodded to the patients waiting on the porch, promising we would be with them shortly. In the front room, all the staff were seated in our customary circle. I took Hotlin aside briefly. She, too, had cried herself to sleep. We hugged each other for a moment in sympathy—and then we prepared to face another day.

Hotlin had recently started a new system for running the staff meeting that we called "spin the pen"—a bit like spinning a bottle except no kisses. Whoever was picked by the spinning pen would lead the morning meeting. We all loved this system, partly because I wasn't necessarily the most skilled at running a meeting. It also added a bit of drama and humor to the start of everyone's day, as people would lean one way or another to avoid getting "got," and then squeal with laughter when the pen chose someone. It turned out that some of our staff, and particularly one of our cleaning ladies, were quite brilliant at running morning meeting. This system was also an extension of our principle of listening to the communities—after all, many of our staff members were now local people. Our drivers were as likely as our doctors to come up with a good idea. The morning meeting provided a structured time when each person's views and ideas were equally valued, and everyone got a chance to talk. Decisions were almost always made by consensus.

This morning, there were a lot of long faces around the room, as several staff members had been present at the meeting the previous day. Farizal gave a brief synopsis for those who hadn't attended, and he shared his plan to visit all the chiefs' houses. A few people suggested that he should visit the resisting leader last, after a lot of other people had (we hoped) already signed, and Farizal agreed. Irawati, who managed the pharmacy, said that her father was ready to sign, so Farizal might want to visit him soon.

As we moved around the circle, the organic farming manager said that ten people were now working in the garden to pay the medical bill for a baby who died last week. The family members had told him how grateful they were that the clinic had tried so hard to save the baby's life and with so much love. The mother told him that she considered their work in the garden as a way the whole family could honor the baby's life, and they were so grateful to be able to pay in this way. The farming manager finished by saying that, with their help, he had harvested eggplants, long beans, and *sawi* (similar to bok choy). Some would be sold in the market, and the rest would be part of our staff lunches during the next few days. In a strange way, I felt that we would also be honoring the baby as we ate the food her family had harvested.

Yuni spoke next; she was Farizal's wife, and she was in charge of our community health program. First, she gave some great news: the rate for patients who refused to continue their treatment in our program, our dropout rate, had gone from nearly half to now only 18 percent. But she was confident we could bring it even lower. She also shared that Dardi was almost finished with his year of treatment for his brain tuberculosis, but his community health worker reported that he was still terrified of leaving the house. "We might need a psychiatrist to help him. Kinari, do you think you can ask the Health In Harmony people to look for one?" I nodded and made a note in my notebook. Apparently, being in a dark room for eight years had caused some psychological trauma as well.

We planned to visit Dardi the following Friday, when the clinic was closed to regular patient visits. (Friday was the day for eyeglass distribution, immunizations, meetings, and home visits.) Agus suggested we could coordinate car trips so that Farizal could plan to visit the head of the village in that area then.

At the close of the meeting, Dr. Frans, who was our new doctor who had replaced Romi when he left to do specialty training, and I went to visit our inpatient, Pak Sofian. He had been staying in the clinic for the last five days. I could hardly believe it when he sat up in bed to greet us! His family had brought him to the clinic in the back of a rented pickup because he had been unable to walk for more than a month. When he arrived, his abdomen was distended, his legs were swollen, and he was gasping for breath. When Dr. Frans first examined him, he gave me an alarmed glance and asked me to listen, too; his entire chest seemed to be filled with his heart, and the murmurs from faulty valves were audible even before the stethoscope was put on his chest. I couldn't imagine how his life could possibly be saved in our small clinic, but as was sadly common, the family was adamant that they could not afford to be referred to a specialist. Here they could pay with seedlings. Apparently, his illness had begun years before with a bad viral infection, and since then, he had been getting progressively worse. However, the fact that he had never tried any medications was good news; it meant that they might work. His relative youth, at about forty-five, was also in his favor.

Now, on day five, it seemed impossible to me that his heart could already be shrinking, but it appeared on examination that this was so. And the heart murmurs had clearly improved. His family members kept hugging us in gratitude, and Pak Sofian himself proudly showed us that not only could he sit, he could actually stand as well. It was like seeing a man awaken from death. A resurrection before lunch was balm for my injured soul and my 3:00 a.m. despair.

While we had been examining Pak Sofian, our receptionist had registered all the new patients in the waiting room, and the nurses were taking their vitals. The computer system Cam designed used photos of people to match the person to their paper record because most people have just one name and no surname, and they may also be called by several other names or nicknames. About 60 percent of our patients also didn't know their dates of birth, and their guesses could be wildly off.

That morning, as was fairly typical, Dr. Frans saw a case of malaria, another case of tuberculosis, a severe headache from high blood pressure, lung disease from years of cigarettes and wood fires, and a long sliver of bamboo that had been in a patient's foot for several years.

While Dr. Frans worked on the foot, I was pulled away to meet with Pak Anto, the head of the national park office, who had arrived at the clinic with some staff members. Hotlin and I invited them into my "office." (This room was also the lab; the clinic's only microscope was perched on the end of the same table I used as my desk.) As we all squeezed in, balancing cups of tea on our knees, Nurse Wil continued to run blood tests right at my elbow.

Pak Anto related that he had been hearing good stories about our clinic, but he had also heard that our meeting the day before with the village leaders had not gone well. I was always astonished by how quickly news spread here. He affirmed, once again, that if the villages would agree to an incentive system, the national park office would help by collecting data on which villages were logging and which were not.

"Do you think it would be possible to get data every three months?" I asked.

"Oh no, we will provide the data every month. My staff are out in the villages, and they *know* what is going on. We will send the data to you every month. How about if we call villages that are not logging *green* and villages that are logging, or have an active sawmill, *red?*"

We certainly didn't have enough staff to collect all the data about logging status, and I trusted Pak Anto. I thanked him for writing letters of support for grants we were applying for, and his staff thanked us for giving discounted care to the national park employees.

We walked Pak Anto and his deputies to the door. I was grateful for their help, but I was still deeply worried that if we didn't get those signatures from the village heads, the whole plan would fall flat on its face. Maybe I had been infected by the fatalism characteristic of so many Indonesians, who often seemed to believe that change was impossible—and that it wasn't even worth trying. I could understand why they felt this way, after three hundred years of colonial rule followed by sixty years of dictators. But I couldn't give up—not yet. Like Pak Sofian, maybe our work could be resurrected.

Dr. Frans still had about ten patients to see. Five others had been sent home with appointments for the following day. We obviously needed another doctor—and we were happy to have found Dr. Yeni, who would arrive next month to stay for a full year to do her year of required government service. The local department of health had just approved our clinic as a site where doctors could fulfill this requirement. In this system, doctors fresh from medical school are often sent into some remote area where, unlike in our clinic, they are on their own with no one to teach them. Dr. Irene describes the system as "a trial by fire that most doctors don't survive." Some doctors are so traumatized, they simply leave their posts. I could understand; fresh out of medical school, I knew *nothing,* and I counted on experienced physicians to learn from. It would have been horrifying to try to figure out what to do all on my own. Yet I also had compassion for the government that needed to provide care to remote areas.

While most of the doctors would stay with us for only a year before going on to do specialty training, I was actually glad for this because we would end up training more doctors who would improve care throughout Indonesia. And sadly, that care really did need improving in some cases, as was illustrated by our next patient, who had returned unexpectedly after we sent him to the nearest hospital in Ketapang.

The patient had been in a motorcycle accident and gone first to the *puskesmas* (government health center) before coming to see us. The big wound on the side of his right knee was still oozing blood; the doctor at the government clinic had simply thrown a few big sutures in the muscles to hold it more or less together. After examining the leg closely, we concurred with the *puskesmas* physician's diagnosis that he probably had a ruptured tendon, and in the end, we decided that he would benefit from general anesthesia and an operating room, neither of which we had. Without a proper repair, I worried he wouldn't be able to walk normally. So we sent him off to our referral hospital in Ketapang, along with a detailed medical note. But now, for some reason, he was back just a week later.

Our nurse Clara—whose fine motor skills and intelligence would have made her an excellent surgeon—was working to pull the stitches out of his hugely swollen and infected leg. The reopened wound was not only draining pus, it had also yielded fragments of asphalt that Clara showed us, horrified. The providers at the hospital had simply sewn up the skin without even cleaning the wound. He had lain in that hospital for a week, getting more and more infected, until finally the family decided to leave—"against medical advice"—and bring him back to ASRI. Hotlin stood at the foot of the bed listening to this story, and then she almost yelled at Frans and me, "I don't care what you have to do, you do not refer a patient to Ketapang unless you absolutely have to! Look at the horrible care they gave him, and now he may never walk well again!"

We had already had many bad experiences with our referral hospital in Ketapang, but this case was truly the last straw. I knew there were many reasons that this might have happened, including inadequate resources, training, and system support. Nevertheless, it was clear that most of our patients would be better off if we took care of them ourselves and hoped for the best.

The main philosophy I tried to instill at ASRI was to love the patients as though they were family members and have compassion for them. When doctors and nurses truly care about their patients, that is also medicine—and from this caring flows the desire to keep learning and a willingness to incorporate new skills. Of course, I also worked on the technical skills, but these for me were secondary. I sometimes joked with the younger doctors (Dr. Irene also had me teaching a class of about ten young physicians, for a few months each year) that there is really only one thing a doctor needs to know: Is the patient *"sick,* or *not sick"*? In other words, which ones do you need to seriously worry about, and who just has something run-of-the-mill? Of course, you also have to decide which *kind* of "sick" they are, what to do about it, and how to look up information you don't know. But this flows from first knowing which patients are of real concern.

Thinking about this, I joined Dr. Frans to see the final patient of the day. She was a kindly looking woman, perhaps in her fifties, who looked unhealthily thin. My senses immediately prickled: *sick.* We sat down, and she began to talk. "I've had blood in my stool. I have a lump in my belly. I'm starting to get headaches. And now, when I walk far, I'm short of breath. I've also been losing a lot of weight." We hardly needed to hear more. And, indeed, on exam we found a hard, solid mass in her abdomen, enlarged lymph nodes everywhere we looked, and noises in her lungs indicating disease there as well.

I looked questioningly at Frans. He knew exactly what I was asking him, since we'd been having a running debate about this. Like most Indonesian physicians, he did not believe in telling patients about a fatal diagnosis, because he felt they would lose hope and therefore die sooner. I understood how critical hope was in all aspects of life, and I could empathize with Frans's desire not to disrupt that. (I, too, felt like I needed an infusion of hope after that horrible meeting the day before.) But I also knew that knowing the truth could allow one to make informed choices about moving forward.

Frans had recently told me about an article he had read from Japan, where the medical culture, similarly, refuses to tell patients their diagnoses—but 90 percent of Japanese people interviewed on the street

Journeying upriver in Gunung Palung National Park toward the Cabang Panti research station. (*Courtesy of James Holland Jones*)

Kinari at Cabang Panti research station in Gunung Palung National Park. (*Courtesy of Cam Webb*)

Candlelit dinners every night at the research station. (*Courtesy of James Holland Jones*)

The hut Kinari lived in when studying orangutans in 1993. (*Courtesy of the author*)

Exploring Native American cliff dwellings after Kinari and Cam's wedding in 1995. (*Courtesy of Cam Webb*)

A radical listening meeting in Sukadana in 2007. (*Courtesy of the author*)

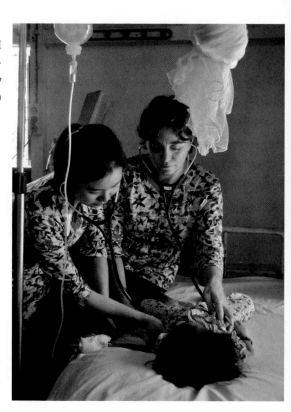

Dr. Vina (one of the ASRI doctors), left, and Kinari examining a patient. (*Courtesy of Ana Sofia Amieva-Wang*)

Dr. Hotlin Ompusunggu in her ASRI uniform. (*Courtesy of Chris Beauchamp*)

Dardi's mother in the doorway of the home she shared with Dardi. (*Courtesy of the author*)

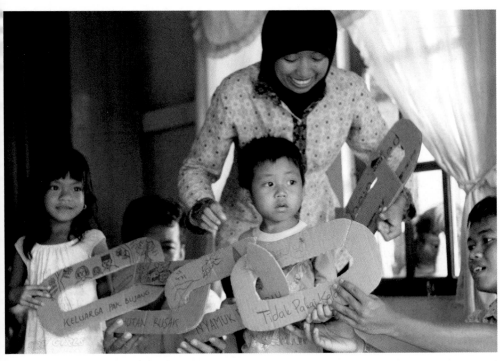

Etty Rahmawati leading a demonstration of the linked causes between deforestation, lack of health care, and malaria. (*Courtesy of Ted Ullrich*)

People exchanging seedlings for malaria-preventing bed nets. (*Courtesy of Ted Ullrich*)

A widow who received one goat from the ASRI clinic now with her multiple goats. (*Courtesy of Katherine Homes*)

Hotlin receiving the prestigious Whitley Award from Princess Anne in London. (*Courtesy of Whitley Awards*)

The first time an ASRI kid saw an orangutan at the Tanjung Puting rehabilitation location. (*Courtesy of Ana Sofia Amieva-Wang*)

A radical listening session with Hotlin and Kinari, listening to a group of community members in Central Kalimantan. (*Courtesy of Hotlin Ompusunggu*)

The ASRI medical center with Gunung Palung National Park behind. (*Courtesy of Oka Nurlaila*)

said they would want to know, no matter how bad the news was. Still, he was not convinced.

In response to my silent question, Frans shook his head. "I just can't do it."

"Do you mind if I do?"

"Go ahead," he muttered, not looking up from the chart.

I turned to this sweet woman and asked, as gently as I could, in Indonesian, "If I had some news for you about your illness, would you want to know?"

"Doctor, that is why I came to this clinic! I heard doctors here tell the truth. I'm dying, aren't I?"

I looked at her lovingly, and after a deep breath, I simply said, "Yes." I reached out to hold her hands as she burst into tears.

"Thank you, Doctor, thank you! I knew it! I just knew it. I have been to see so many doctors, but all they would tell me was that I had an 'irritated stomach'—but I just *knew* it was serious. The problem is, I had to decide whether to call my family to come from Java. I am so grateful that you told me, because now I will ask them to come as soon as they can. I want to see them so much before I die, and there are many things that I need to tell them."

We talked for a few more minutes, asking and answering questions. It was metastatic colon cancer, I told her, and it was indeed too late for treatment. She got up, grasped both our hands in gratitude, and gave me a big hug. "Thank you, Doctor, for giving me the chance to say goodbye to my family!"

After she was gone, Frans looked at me with moist eyes. "Okay, I am convinced. If they want to know, I will tell my patients the truth."

OVER THE NEXT MONTH, FARIZAL visited almost every village head (as promised), to ask them to sign the agreement to work with us. He often slept overnight in a village to catch the headman at home the next morning. Farizal visited those he knew would be easy first, and sure enough, he rapidly got eighteen out of twenty-one signatures. Even just with these, we could go forward, but now the three hard ones were left.

The first of the three was a failure. The headman of Pampang Harapan refused to sign because he had run on a platform that he would work to change the borders of the national park. The second "difficult" headman was in Matan. Again, he refused, despite the fact that every time we went to Matan, people expressed a strong desire to have their leader approve the agreement. We suspected he was also involved in the logging. Finally, only the village of Penjalaan was left. Hotlin and I accompanied Farizal for this critical visit. But when we got to his house, we found him looking like a deflated balloon. He quietly asked us how many village leaders had already signed. A still nervous Farizal told him that eighteen had agreed. The headman hung his head, sighed, pulled out his pen, and quickly scribbled his name. Apparently, he had decided he could not deny his village discounted health care, even if it might hurt his business. And indeed, a week later, we heard that he had stopped hiring people to log the forest and had even closed his sawmill and opened a small store. And soon after that, he actually stopped by the clinic to say hello and have a cup of tea with us. We welcomed him warmly, happy to be on a much better footing.

By July, a year after we opened, the pharmacy now boasted a new red-green pricing system, which, along with the mobile clinics, would hopefully help reduce the cost and burden of health care on patients and the community. I just needed to make sure that we would continue to have enough money to provide these thank-yous from the world community, and that meant keeping close track of every expense. I think of the accounting as where the rubber meets the road of a nonprofit.

We had funds from the U.S. Fish and Wildlife Service to do reforestation, and I was using the seedling procurement funds to "buy" seedlings from our patients and transfer that money to pay for their medications. We had explained we would do this in the grant application, and they loved that an additional benefit of reforestation would be health care.

Even though barter was an option, most of our patients were still choosing to pay with cash because our prices were low (the green price was about the cost of a bowl of soup). However, the more expensive bills tended to be paid with noncash options, and we often heard what a sense of relief this alternative gave patients. Overall, we were bringing in about a third of the cost of care.

We were getting help understanding the economic situation of many of our patients, thanks to a Yale medical student, Mei Elansary, who was conducting patient surveys. So far, she had learned that 69 percent had had to choose between food and medicine in the past. The best estimate of household income in our area had been done by the World Health Organization in 2003 and showed an income of just thirteen dollars per month. It had likely increased since then, but probably not much. Knowing that people might have to log or not eat to pay for medicines made us even more careful to dispense medicines only when they were absolutely necessary. The docs I had worked with were amazed to discover that patients were often content to return home with just an explanation and advice about things like diet and exercise. Anxiety or psychosomatic symptoms could often improve simply with the doctor's explanation that the patient was well. My father—who, like my mother, has a Ph.D. in psychology—calls it being "cured by a diagnosis." I have seen that truth affirmed thousands of times and even experienced it in my own life.

Mei was also asking about the cost of care before people came to our clinic. Last week, she had interviewed a father who brought in his daughter with a horrible skin condition. Dr. Yeni was pretty sure what it was but called me in for a consultation. I agreed that it was scabies— just a really extreme case. (Scabies are highly infectious tiny mites that burrow into the skin and cause severe itching. In Indonesian, this disease is called *kuties,* which may be where the word *cooties* comes from.) Dr. Yeni prescribed two dollars' worth of medicine to treat her and the entire family. They couldn't believe the bill and chose to pay with cash.

Before going home, they were randomly selected to be in the survey.

What Mei found was extremely distressing: they had already spent five hundred dollars trying to cure her, consulting doctors, nurses, a pharmacy, and a *dukun* (traditional healer). This is a vast sum of money locally and would have almost certainly required logging to get. They got medicines from the medical providers that not only didn't make her better—they likely made her worse, and they stopped going to the *dukun* only when he started itching, too.

Dr. Yeni poked her head in my office door and told me she wanted me to see something. I was relieved to be able to leave the accounting. In the waiting room, a distinguished-looking older man with a scraggly white beard and Islamic prayer cap was standing in front of everyone, giving a speech that I just caught the tail end of.

"This clinic is so amazing! I am so glad it is here. Just like when you come to see me and you pay with chickens, here you can also pay with barter! I'm so happy that I can pay with seedlings! I'll be back in a few months when they are big enough to pay my bill. Just know that this place has my full support."

When I asked what was up, Dr. Yeni explained that he was the *dukun* who had gotten scabies from the little girl she'd cared for the week before. She had just treated him as well!

A few hours later, Mei and I sat together over lunch. She was bubbling with excitement about her interview with the traditional healer, who had also been randomly selected for an interview. This *dukun* had explained a phrase that Mei had been hearing over and over again in the interviews: *terkena badi*. I had never heard this term, and it soon became clear why patients would not mention it to me. The word *terkena* means to be "struck" unexpectedly, and the phrase *terkena badi* (sometimes shortened to *badi*) means to be "struck by the revenge of a spirit being." According to the *dukun*, this could happen by disrupting the natural environment in some inappropriate way. Mei pulled out her notes and read me the translation her translator had done from the recording: *"A person who cuts down a tree can get* terkena badi. *Trees have powerful spirits. The spirits are created by Allah, but they are the ones who protect the trees. The earth (ground) also has spirits. Water has spirits. . . . We cannot cut down trees arbitrarily, because there are things that cannot be seen. People who cut down trees can get sick, they can even die."*

"So this explains why two-thirds of our patients say they have consulted a *dukun* first," I mused. "Maybe they only come to see a medical doctor if the treatment for *badi* didn't work." Then I smiled. "Maybe in the future, we will start getting referrals from him. There is much that traditional medicine is great for—but some things, like scabies and tuberculosis, just can't be healed. Apparently, he recognizes that." Clearly, each medical system had their specialties, and we would have to work closely with the traditional healers.

With this new understanding, I had to laugh to myself when one of our next patients was back for a second visit with what I now knew were classic symptoms of *badi*: one cold foot and pain in her stomach that radiated all the way to the top of her head. We had clearly missed the diagnosis the first time! I decided to give it a try and ask her if she thought she had *badi*. Her narrowed-eye sidelong stare at me and drawn-out denial told me that is exactly what she thought she had. When curing her with a diagnosis failed, Frans and I decided she would be our first referral back to the *dukun* since neither of us were experts in curing spirits.

I walked back toward my office and saw Farizal stapling photocopies of the signed agreements to give to the national park office. (He had counted the hours that organizing the meetings, conducting them, and going to each head of the villages to get individual signatures had taken, and he'd come up with four hundred.) I sat down next to him and asked if he knew about *badi*. "Of course," he said. "I talk about it often when I talk about the importance of protecting the forest. Everyone knows you can get sick if you log the forest. People say if you log the forest, it pushes the diseases down from the mountain into the villages."

Given there are types of malaria that live in macaques but can affect humans in these forests and other unknown zoonosis, there was probably epidemiologic as well as spiritual truth to this (note that Ebola, HIV, SARS, and MERS are all diseases that originally came from humans invading into wild ecosystems). Farizal and I talked about how these communities had been trapped in a downward spiral, where lack of access to good care led to more logging, which led to more disease. If we could improve the quality of care, allow people to pay with barter, help with transportation costs through the mobile clinics and ambulance, *and* give them discounts for protecting the forest, we might be able to decrease the pressure on the forest and improve the health of the communities. If we could protect their health, it might also prevent new diseases from the forest from spreading throughout the world. While most people globally were totally unaware of it, the health and well-being of people living around a forest in Borneo really did matter to everyone else's well-being.

(13)

Connecting with the
Outside World

LATE 2008–EARLY 2009

S ometimes I liked to joke that what was needed in one place in
the world always seemed to be in excess in another. In 2008, one
of my former professors at Yale, Michele Barry, came to visit.
Michele was upset to see how much I was doing every day—teaching,
accounting, grant writing, grant reporting, overseeing the conservation
work, budgeting, working on the pharmacy list, to name just part of it
all. "Kinari, you obviously need help! You are doing the work of ten
people. At least I can get you volunteer doctors to help with some of the
teaching." Soon we had a steady stream of physicians coming to help
teach.

Health In Harmony was also able to bring in a very specific expertise
we needed: psychiatry advice. Dardi was experiencing severe agorapho-
bia after so many years in a dark room. Volunteer Dr. Peter Mayland
arrived and was soon working with Dr. Made (pronounced "MAH-
day," which means "second-born" in Balinese), who was one of the doctors who had replaced Dr. Frans. Dr. Made learned from Peter how to
lovingly and calmly help Dardi. After their first session, Peter told me

that in all his years of working with patients, he had never met anyone with more fear. The goal was to slowly desensitize Dardi until he could actually leave the house. At the first session, Dardi managed about one whole minute of sitting near the front door of the house, supported closely by Peter's presence and shakingly gripping the edge of the door. At the end of their fourth and last session, Dardi could sit in the fully open doorway for a few minutes at a time. Peter taught Made how to continue the work weekly. One day, after about another month, Dardi stood up and said he wanted to *walk to the road*. He managed to do this, firmly clasping Made's hand. When they got to the car, Dardi smiled a big smile. "Can we go to the beach? I remember going there when I was a little child." In disbelief, Made and our driver took him to the beach, about twenty minutes away. Made and Dardi sat side by side on a log, looking out over the scoop of the bay, watching fishermen bringing in their nets. Dardi still held tightly to Made's hand. After only a few minutes, he said, "This is the best day of my life. Thank you, Dr. Made. But can we go home now?" This was a huge step forward, and the whole team rejoiced that after two years of care, he was taking tentative steps back into life again.

There were also unexpected benefits of starting to connect Sukadana to the outside world. Peter and his wife, Laurel, like all our volunteers, shared housing with our Indonesian staff. They were living with Farizal and his wife, Yuni (our tuberculosis program coordinator). Like most volunteers and staff, they formed a fast friendship—but I was very surprised when I found out Peter and Laurel had decided to pay for Farizal to accomplish his dream and do a master's degree in sociology. This was a hugely positive thing, *and* I would have to find someone to replace him to carry on the conservation work.

Not long after this, my mother and stepfather came to visit, and I decided to go with them for a short vacation in Bali. While there, we looked up someone that my German host sister raved about. The woman's name was Etty Rahmawati, and my host sister knew her from a semester she spent studying Indonesian. Etty had a job in Bali teaching

English, so I invited her to join us for dinner. Etty kept us so entertained, and we were laughing so hard all evening, that we started to get nasty looks from everyone else in the restaurant—apparently, we were having a little too much fun. Or maybe Etty's headscarf had something to do with those looks? Hindu Bali still had hard feelings against Muslims after the Bali bombing in 2004, which for years disrupted the tourist trade that sustains the island's economy. But Etty, like most Indonesian Muslims, is extremely moderate. While she wears a headscarf, prays five times a day, and dreams of going to Mecca for the hajj, she has many close friends from different religions, and she seems to not care what path others might choose to come closer to the Divine. Someone's religion is no reason for her not to love them with her full heart, and in return, she is loved back by almost everyone who meets her.

I was so impressed with her intelligence, her charm, and her ability to capture people's attention that I began scheming that very first night about whether ASRI could hire her to take Farizal's position. Luckily, a few months later, I was able to tempt her away from her prestigious job in Bali to be our new conservation education coordinator. Very quickly, her popularity in the communities outpaced my own. As we rode our bikes around, the shouts of "Mbak, Etty!" became much louder than the "Hallo, Doc!" that I usually got even when in extremely remote places. Soon after arriving, we put her to work helping distribute four thousand mosquito nets that were donated by the Against Malaria Foundation in the communities around the park in exchange for seedlings. In the midst of this effort, we had a surprising visit that ended up bringing in more resources: a PBS film crew had heard about us through one of our donors and wanted to do a story for *NewsHour* with Jim Lehrer.

They were there for only three days, so we tried to cram as much as possible into their time. On their first day, we filmed the clinic, including morning meeting, Etty explaining the red-green system to the waiting patients, and doctors doing rounds. Next, we took them to see the organic farming initiative, which had been thriving for the last six months. We had been able to meet the local need for training in sustainable agriculture quite close to home. The next-door island of Java has a many-thousand-year tradition of regenerative agriculture, and we found an amazing organic farming teacher who had worked at a nonprofit in Java, training street kids in farming skills. Pak Ngalim was reminiscent of my favorite high school biology teacher whose hands-on illustrations brought lessons to life. At our first major three-day training event, one of the things he taught the gathered farmers and loggers was how to make a compost accelerator. They set some water to boil on a little campfire under the tarp we had set up for the training. Then Pak Ngalim sent each group of three off to get a rotten banana trunk, and within minutes, every group returned with a thick stalk balanced on a shoulder. When the sterile water was cool, they mixed in shrimp paste, rice husks, and sugar to make a micronutrient and protein-rich slurry. Into that went cut up pieces of the rotting banana trunks so all the bacteria and fungi breaking down the stalk could flourish. Three days later, it was poured on a pile each group had created of manure, cut green grass, and scraps from lunch. A few months later, voilà, excellent fertilizer nearly for free.

Now, with the film crew, we visited one of the new collaborative organic farms that a group of men and women created after attending Pak Ngalim's organic farm trainings. The head of that cooperative explained the critical importance of organic farming: they had been able to revive land that had compacted and been abandoned. He described each year needing more and more chemical fertilizers until they could simply no longer afford it. He also talked about how much more money they were making from selling vegetables than they had made from growing just rice in the past. Now they were making enough profit that some of their members could shift completely away from logging.

After leaving the organic farmers, we went to one of the twenty-three mosquito net distribution events we had been doing all around the park. Watching, through the camera, so many people completely fixated on Etty's talk, I thanked our lucky stars once again that I had been able to steal her away. As in all our work, ASRI's mosquito net distribution integrated human and environmental well-being. After giving a demonstration of how to use the bed nets, Etty launched into a story to explain why we were asking for tree seedlings in exchange for mosquito nets. The story was of a family who lived near the forest. Some of their neighbors logged, leading to more mosquitoes, which caused the father to get malaria, since he did not have a mosquito net. He continued to worsen, so his wife called together the family, and they made the difficult decision to borrow money despite the very high interest. Sadly, they were too late, and he died before reaching the hospital many hours away. Now his widow had to pay back the debt, and the only way to do that was to get her family members to log the forest.

After relaying this story, Etty looked around at her listeners and asked them whether this story sounded familiar. One woman called out that she once had to pay 150 percent interest per month on a debt, and other people commented that they had personally experienced various parts of this story. Then, using David Werner's creative technique, Etty asked one child on her right to hold up a cardboard cutout of a happy family in front of a house. Then she asked another child on her left to hold a cutout of the grave.

"Can you help me figure out what led from *this* to *this*—what are the links in the chain?" She gestured to the family and then to the grave.

As people called out each reason, Etty put together big cardboard links to form a chain of tree stumps, medical syringes, and money signs.

"Our goal at ASRI is to work together with you to break *every one* of the links in this chain. Can you all give me ideas about how we could break each link? Some of them are things that ASRI may be able to do, but others, only you can do."

Again, people offered ideas. The mosquito nets would help, and they were grateful to have access to a good clinic with ambulance service. People suggested that the village should find alternative work to illegal logging. And they should try to replant the forest that was cut. *Now* they understood why we were asking them to give tree seedlings in return for the mosquito nets.

Etty has a musical, lilting voice, and she knows how to emphasize and pause at just the right moments. I was able to translate for the film crew, and I think the story helped them understand ASRI's work so that they, in turn, could help their viewers begin to see the connections between human and environmental health.

After her talk, I joined the producer, Nikki See, as the film crew videoed people handing over the seedlings they had grown for their mosquito nets. She asked me about the data on deforestation and malaria. I told her that while it wasn't universal, the majority of the research all over the world was showing linkages between logging and *Anopheles* mosquitoes, which carry malaria. One study published in 2006 in *The American Journal of Tropical Medicine and Hygiene,* from the Peruvian Amazon, in which the researchers used their own legs as bait, showed that the biting rate of *Anopheles* was 278 times higher in logged versus unlogged forest. Another study showed that women in Indonesia were likelier to die in pregnancy from malaria if deforestation was happening while they were pregnant.[1] I also shared my own experience (admittedly, unquantified) of being bitten by only a few mosquitoes in the primary forest at Cabang Panti, in comparison with the logged forest behind our house in Sukadana, where the mosquitoes were so dense that I hated to go in. This may be in part because logged forest likely has more standing pools of water, as well as fewer mosquito predators.

Although the film crew wouldn't get to see the planting of the more than fourteen thousand seedlings we had collected, I told them that when the community members gathered in the treeless landscape selected to put them in the ground, it would be equivalent to taking 2,333 cars off the road for a year (although, of course, carbon sequestration is just one tiny part of the myriad ecological, cultural, and spiritual benefits of these forests and not our primary purpose). The area where we would start planting would expand a corridor of forest that was all that remained after horrible logging and fires. Dark ghosts of burned

---

1 Chakrabarti, Averi. "Deforestation, Malaria and Infant Mortality in Indonesia," SSRN [2018]: https://ssrn.com/abstract=3257339

trees still loomed here and there over the planting site; but someday, we hoped, they would be replaced by new, towering giants of living trees.

Cam had designed an experimental method for testing the best planting methods, and I explained that one of the trials was to see if cardboard rings around the seedlings would help keep the grass down. Since we would soon have as many seedlings as we could comfortably plant, other communities were helping cut cardboard rings in return for their mosquito nets. We had been dumping big loads of uncut cardboard in each community, bought from the trash pickers in the city.

We had a Canadian volunteer, Andrew McDonald, who had been helping with all the reforestation efforts. I told the film crew a funny thing he had told me. He called the cardboard rings the perfect *asri*. When I was confused, he explained that he called an *asri* something that perfectly balances human and environmental well-being at each stage along the way. The trash pickers got work; hundreds of community members cut the rings in exchange for mosquito nets that would keep them healthier; the rings would be used to regrow rainforest, which would help make the villages healthier; and even the cardboard wouldn't end up in a landfill but would rot into the soil and give nutrients to the seedlings.

I told the producer we had gotten one complaint, though. The village head of Laman Satong—the village where we were doing the reforestation—had mock grumbled to me: "Since you gave us mosquito nets, everyone in my village has become very lazy. Now they sleep in in the morning and don't start work right away." He had said this with such a big smile that I figured he was probably one of those newly lazy ones.

The last part of the filming was an interview with me. I wanted them to interview Hotlin, too, as she was so critical to the work—and because she was much more eloquent in front of a camera—but they insisted that it would have to be me. This same thing had happened when an Indonesian newspaper, *Kompas,* had come to do a back-page story on our work. The whole thing with PBS was so nerve-racking that, I am ashamed to say, I acquiesced. But that meant that the true story of our work being done primarily by Indonesians was not getting out. I was keenly aware that journalists have incredible power, but I hadn't yet learned how to navigate them, convey nuances, and make sure they got the key points.

In the end, the interview went off fairly well. We were glad they could help us spread the word about the importance of these forests for the well-being of the whole planet and how the lungs of the earth wouldn't survive if the people around those forests weren't healthy as well. However, they did get one critical point wrong: they thought we denied care to logging villages, which, of course, we would never do. They also failed to truly grasp that what we were doing was something fundamentally different. We weren't coming up with the solutions to save rainforest—the communities were. Yet this radical story was so far from most reporters' preconceived assumptions that it was proving very difficult to get across.

14

Pressure Cooker

SUKADANA, WEST KALIMANTAN:
OCTOBER 2009

In 2009, financial and emotional stress began to stretch me to the limit. I traveled twice to Europe and twice to the United States in addition to all the work at ASRI. The disconnect between the small Bornean village and the fancy parties in the global north was also psychologically straining. I experienced the entire spectrum of human wealth in the thirty-hour time warp of an international flight: from one-room huts, where ten people lived on a diet of essentially rice, to fabulous apartments with skyline views, serving up champagne and salmon pâté.

Despite the fundraising trips, including one Hotlin did to the States, our finances were quite strained. In forgone income and frequently covering expenses, Cam and I were the biggest donors to the program, and that couldn't continue forever. There were many months when the nonprofit's bank account had just enough to pay salaries that month. I mostly kept this to myself, as I didn't want the staff to know how precarious their jobs were—but this worry was especially hard on me. In the middle of a Health In Harmony board meeting in the States, I broke down and cried. It just wasn't tenable anymore for me to help Toni write grants, do the accounting, manage the website, write newsletters, deal with volunteer inquiries, fundraise—*and* oversee the clinic. Most of the time, I felt like I was leaning forward at full tilt. Brita Johnson was also overwhelmed. She was officially Health In Harmony's executive director, but she was also serving as the organization's secretary, development director, communications director, and anything else we might need—and all this was happening out of her kitchen. Brita did a great job, given the limitations, but she obviously needed more help than our many dedicated and hardworking volunteers could provide. The board agreed we would have to hire more staff, especially someone to help with fundraising in the States, but again, this felt like a big risk. I was doing my best to come up with creative fundraising strategies, but we needed someone focused on this.

One of the new strategies simultaneously raised money and solved a new problem. After three large trainings in organic farming techniques, there was now a shortage of manure. There had previously been no price for manure in any of the villages, since no one was aware that it could be used for anything. But now, there was not enough to meet the new demand, and once people learned they could make fertilizer from it, they even began to hoard it and were unwilling to sell it for any price. So we started a new program called Goats-for-Widows, where we gave a mated pair of goats to women whose husbands had either died or abandoned them. (The Indonesian term *widow* can mean either, and the community wanted these particularly disadvantaged women to be helped.) "Selling" these goats to donors was proving to be an effective fundraising technique, as people gave them as presents and were happy to help some of the poorest members of our communities, who were often the sole supporters of large families. The idea was that each woman would pay forward one baby goat to another widow, and also give a few bags of goat manure to the farmers we had trained. After that, they could sell additional kid goats and bags of manure for a small, steady income that allowed them to afford important "extras," such as sending children or grandchildren to school.

We also managed to pay for our ambulance through a creative fundraising campaign, where generous donors sponsored each seat belt, engine part, door, winch, and bumper. We couldn't just *buy* a four-wheel-drive ambulance, however; a truck had to be specially adapted and then shipped from Sulawesi to Borneo. They had fitted a cab over the truck bed and installed shelving, a bench seat, and the rollers for the stretcher. Then the whole thing was painted to look like an ambulance, with lights added to the top. It wasn't the best ambulance ever, but it worked, and it could manage all the horrible roads.

The second big delivery was the satellite internet. The staff loved it, and some were even coming in at night to take online courses, but the price was outrageous. A meal and a drink in Sukadana could cost about two dollars, but the internet tariff was four hundred dollars per month. Having better internet eased communication with supporters in the United States, helping me raise money—but it added other difficulties. Cam had convinced Harvard to pay for half of it (because he needed it, too), but he also had to fix it nearly daily, which was a huge strain on him—and, by extension, on our marriage.

During this year, we had an American expat on our mailing list reach out to ask if she could visit. I had met her in Jakarta through friends when she was working as a project evaluator for USAID in Indonesia. At the end of her visit, she confessed that she had come because she figured we *must* be lying about our results. She had felt certain that no program could accomplish this much in such a short period of time in Indonesia. But her final words to me were: "You *were* lying—you haven't shared half of what you all are doing!"

She was right; I hadn't been sharing everything, because I worried that our supporters would think this pace was unsustainable—and they would have been right. In addition to distributing the mosquito nets, we were building a conservation office, and Americares had sent us thousands of doses of anti-worm medication that would expire within a few months. We had been forced to quickly design a community-wide worm eradication plan—including teaching good handwashing techniques, since handwashing is apparently as effective as medications. In all these cases, we could get the funds to do the thing or get the supplies, but none of these programs wanted to help pay for staff to actually do the work. I understood why that was the case, but it was frustrating.

On top of all of this, we also held our first cataract surgery event in partnership with the local government, giving the gift of sight to seventy-seven blind people. Luckily, the volunteer ophthalmologists (one American and one Indonesian) fundraised for the event, but again, those funds only covered all the equipment, not our time. It took weeks to screen all the patients, days to convert our clinic to temporary operating rooms (with plastic sheeting), and then two intense days from dawn to past midnight making sure that every time a patient came off the operating table, another was perfectly prepped and ready to lie down. It felt worth it, though, when we got to remove the bandages from patients' eyes on the second day and witness them seeing for the first time in many years. One man gasped when he saw me and told me I was the most beautiful woman he had ever seen. I laughed and told him to just look around, and I was sure I would lose that status. Each person had been led into the room by family members. They all had a bandage on and one blind eye, usually with a distinctly white pupil. But after their bandages were removed, they just stood up and walked out of the room to greet their family members. One old woman swung her eight-year-old granddaughter into her arms with joy. Though this girl cared for her every day, the grandmother had never actually seen her. They rejoiced that now she would not have to stay home and would be able to go to school.

But all was not rosy. That second night, I went into the yard in front of the clinic and called our driver to pick up the food for the staff. When he finally answered, he said he couldn't go—he was two hours away interviewing for another job.

"How could you not tell me you were interviewing for another job? How dare you leave us in the lurch! Just when we need you the most! And on this day, of all days?" I yelled at him.

His indignation that I would dare to be mad at him made me even madder, and I hung up on him. Now Hotlin or I would have to pick up the food and drive the surgeons to their hotel that night. Fuming, I turned around to see many of our staff staring at me in open-mouthed shock. All my anger fell away, and I felt terrible. I knew the rules in Indonesia about not expressing anger, and I had blown it. Clearly, the stress was wearing on me.

When I showed up the next morning at the clinic, I discovered that our driver had dropped off the car keys before I arrived and quit. I tried calling him many times, but he would not pick up. Finally, the following day, I went to his house and sat down on the broken couch with rusty springs sticking through the plastic. His wife came in crying, carrying their few-month-old baby, whom I had helped deliver and told me he left and she didn't know when he would be back. I asked to hold the baby and just sat there rocking her and feeling guilty. How much of him interviewing somewhere else was my fault because I was over-working the staff? And had he now left his family because I yelled at him? *When would I learn?*

For two whole months, he didn't return, and during this time, I had trouble sleeping. I felt terrible about the whole incident, but personal struggles were also agonizing. Cam and I had been trying to decide whether we should move forward with adopting. It was extremely difficult in Indonesia for foreigners to adopt a child, but after a year of nearly full-time work, Toni (who also was having trouble conceiving) was nearly done with the paperwork for little Thomas. I had begun collecting the needed letters, but Cam was very uncertain about it, and so was I. Should I take all the love that I was giving to a whole community and focus it on one child? We could also potentially go the very expensive medical route, but was that a wise way to spend resources in the world?

Then there was something else that was really worrying me. I feared there was a fundamental flaw in a key aspect of our work. It was looking like we couldn't trust the data we were getting from the national park staff about the logging status of the communities. Their reports didn't always agree with what we ourselves could see and what the community members were telling us. I trusted Pak Anto, but we suspected that some of his staff on the ground might have reason to fudge the logging status in the areas they oversaw. My anger at them dissipated a little when I discovered that at least one of the park staff had a child who was undergoing extensive (and expensive) medical care—and even worse, the diagnosis was almost certainly incorrect. To make matters worse, Pak Anto himself was being moved to another park. That had been the last straw, and we made the decision to call an evaluation meeting with all the community leaders to discuss what to do.

I was acutely aware that the monitoring system was the only part of the program that *wasn't* designed by the communities. I was frustrated with myself that I hadn't trusted them in the first place to tell us what the best system would be to determine which villages were logging and which weren't. I claimed to believe in radical listening, but clearly, I hadn't fully embodied that.

On the day of the meeting, Hotlin and I were seated next to each other in the circle of chairs. All the other chairs were occupied by the twenty-one male village leaders. The sound of chain saws could be heard in the distance. We started off by apologizing and asking for their help. After a few minutes of slightly irritated humphing and agreeing that the red-green data was incorrect, they settled into finding a solution.

"Your first problem is you should be measuring logging at the *dusun* level, not the *desa*," one leader said, to wide agreement from the group. This man was referring to the government administration units. These men, who had signed the agreements, were all *desa* leaders (about three thousand people), and each one of the *desa* was made up of about three *dusun,* which roughly correlated to a village (about one thousand people).

Another leader jumped in. "This is totally true. In a *dusun,* everyone knows everyone else, and they can probably push a logger to stop—but at the *desa* level, we don't all know each other.

"People really care about those discounts, so I think the logging would stop much more quickly if it's on a *dusun* level. But you will need one person in each village to help you do the monitoring who speaks that language and who knows every logger because they grew up together. They can also make sure everyone knows the village's status."

This idea again resonated with the group, and it was quickly agreed that we needed about thirty part-time staff living in each *dusun* who would keep track of the logging activity, try to convince the loggers to stop, organize training events, and coordinate with an ASRI staff member who wasn't local and who would make sure their reports were honest.

Hotlin was excited about the idea, but I felt uncertain how it would work and worried about where we would find the money for so many more staff, even at low salaries. I just wished we weren't so dependent on donations and mostly on my own energy to make that happen. I wasn't the only one worried about this; donors and grantors were often asking how we were going to make the program *"sustainable."* In reality, these communities were giving much more than they were receiving, but we hadn't figured out yet how to put a monetary value on protecting the future of the planet. Hotlin and I had continued to work on the possibility of carbon funding. But after two full years of studying various carbon offsetting mechanisms, even doing a feasibility study in 2009, and creating a consortium of potential stakeholders in the region, we had finally given up. The high cost of verifying how much carbon was in the forest (done by literally measuring trees) in addition to all the political uncertainties and the high tax the government would take, made the whole thing seem nearly impossible. I had heard of brave souls in other parts of the world trying it, but I hadn't yet heard of a real success.

What I wished for was a carbon tax—or, as Canada calls it more correctly, a fee on pollution—and to have some of that money dedicated to communities so that they could save rainforests. If a fee on pollution was imposed, then people would naturally want to buy products that damaged the earth less (now it's essentially impossible as a consumer to make choices based on pollution levels). Sadly, though, the implementation of a pollution fee didn't look politically popular anywhere because of typical short-term thinking. So even though our work had the potential to vastly benefit the world, it seemed like magical thinking to see a way around relying on donors.

Then in the midst of all this stress, Cam got some very good news. He had just been awarded a prestigious terrestrial ecology award from Inter-Research for the best young scientist in his field; one of his papers had in fact created an entirely new field in ecology, called *phylodiversity*. I had already agreed to help Cam teach his eight-week Harvard summer course called the Biodiversity of Borneo, and we would have to squeeze in this trip to receive the award in Germany before going on to Malaysian Borneo. There we would meet up with the ten Harvard students and ten students from across Southeast Asia (Cam had said he would teach the course only if they charged each Harvard student twice as much to cover the cost of a Southeast Asian student).

After the long trip to Germany, our second trip to Europe that year after going in the winter for my sister's wedding and to see Cam's parents, we arrived in Kota Kinabalu and immediately dove into organizing daily lectures, visiting the vast underground caves of Mulu, trekking in two rainforests, snorkeling on a coral reef (which involved teaching some of the students to swim in their new burkinis), and finally preparing to climb Mount Kinabalu. At the bottom of that thirteen-thousand-foot mountain, I wasn't at all sure I had the energy to do it. One of the students from Papua New Guinea swung the strap of her bag onto her forehead and powered away up the mountain. There was no chance I would keep up with her. I loved seeing how the Southeast Asian students discovered that they could compete in all ways with the Harvard students. They, too, were discovering that there isn't that much variability in human intelligence.

After getting home to Sukadana, I was bone tired and eager to finally have some time to relax with Cam. But that is not what happened. Ten years earlier, Cam had agreed to organize a giant conference in Indonesia in 2010, for the Association for Tropical Biology and Conservation. He thought it would be an excellent opportunity to enable more Indonesians to participate in a world-class conference, but once our trip was over, the terror of that looming nine-hundred-person meeting had him waking in the middle of the night in a cold sweat. He'd already been working intensely for more than a year on it, but now the pace became punishing. He would wake at 3:00 or 4:00 a.m.; if he came to the dinner table, his brain did not accompany him; and evening swims or Sunday afternoon DVDs became a thing of the past. He exuded stress and it only ramped up the difficulties in my work. I found myself angry at him. I felt married to a ball of anxiety, not a human being. Finally, I just said to myself, *Kinari, give up. You won't have a husband for at least a year. Don't even try.*

Then the pressure cooker exploded. But surprisingly, it wasn't Cam or I that exploded—at least not just then. It was Hotlin.

15

Systemic Wounds

LATE 2009–EARLY 2010

In Indonesia, it is considered a sign of maturity to be able to tightly control your emotions. So when the lid does come off, the result may be more explosive than Westerners are used to. Hotlin's expression had become increasingly bitter over the preceding weeks, and she was turning more inward. I tried to draw her out on several occasions, but she refused to talk with me. I planned a special meeting with the whole staff to review all our achievements, thinking that it might help to buoy everyone's morale, and especially Hotlin's. But after I gave the presentation, she started raging at me in front of everyone, "You have destroyed my life!" I sat dumbfounded. She continued: "I hate you! I trusted you. I put my faith in you. You brought me here to this remote village where no one knows or respects my culture—where I give *everything* to help people, and then they are ungrateful and even get angry at me!"

I tried to interject, but she was having none of it. "If it wasn't for you, I'd probably be married now! I don't have savings or a house or anything! I'm stuck here in a village in the middle of nowhere where no one appreciates me. No one loves me. How could you do this to me? I hate you! I'm getting on the boat and leaving tomorrow!"

She stood and marched right out of the clinic, and I and the entire staff just sat there with our jaws dropped. Then I slumped in my chair and hung my head. Staff leaped up and patted me on the back and insisted that I was not to blame. But I knew there was some truth to what she'd said. All the ASRI employees were getting lower wages than they might have earned elsewhere (though by now with steady increases). While I had never asked Hotlin to work as hard as she did, I had set a bad example of overwork, and I'm sure my stress was apparent to all. It was also true that she was unmarried and in her midthirties. Her chances of finding the right man in a remote village in Borneo were indeed low. I cared about her so much as a friend—but maybe I hadn't been showing it enough.

I stood up from the meeting to go after her despite the protests of some of the staff.

I found Hotlin crying in the "girls' house," sitting on the wicker couch, with a few of her clothes already thrown into a suitcase on the floor. I sat next to her and put my hand on her knee. She would not look up. Only with much prompting did she begin to tell me all the things I had done wrong.

"When the video crew was here, I felt like you didn't even want me there. No one recognizes what I have given for this. I know you tried to include me when *Kompas* came to do a story, but they wouldn't even listen to me. And when you kept trying to make the reporter take a picture of us together, he kept cutting me out, and the picture he ended up using was the only one of just you. I know I shouldn't care about these things, but I do. Did my mother give up her honorable funeral for this? So that I could be miserable here?"

Hotlin comes from an ethnic group in Sumatra called the Batak who are famous for their elaborate funerals. But these fancy ceremonies can only happen for elders who are considered to have lived a good and successful life—and one of the criteria is that all one's children are married.

When it became clear that Hotlin's mother did not have much longer to live, she demanded that Hotlin quickly accept an arranged marriage. Hotlin did not feel she could refuse this dying request, but she could not hide her misery when the choices were offered. At the last minute, her mother relented and decided she would not force her daughter to marry a man she did not love. She died believing that she would have a "dishonorable" funeral.

I knew that this had been a critical moment in Hotlin's life and that Hotlin missed her mother terribly.

"Oh, Hotlin, I am so sorry!" But I felt somewhat defensive and at first tried to explain myself. "It isn't true that I didn't want you there when the *NewsHour* people came. But I was nervous about the whole thing, and they wanted to tell the story of the American doctor."

Then I just stopped and said, "I know it isn't fair—it's even racist. I hear you that what you have given is even harder, if it isn't recognized." I knew I was not blameless. I gave a lot, but I also asked a lot, and I wasn't always sensitive enough to the needs of others (or to my own, for that matter). I put my arm around her while she cried for a while.

"It's possible that it will work out. That's what happened with your mother's funeral. She chose to just love you, and then the tribal elders ended up giving her an honorable funeral anyway. They couldn't deny she had lived a good and faithful life. And you are like her. You are so kind and ethical."

Slowly, her body began to relax, and I added, "I know this has been a sacrifice for you. I also know from everything you have told me about your mother that she would have been so proud of you."

By this point, my eyes were also moist. "I will support you to leave. I will do whatever I can to make your life better, and you will have my blessing. But you must also know that I don't want you to leave. I don't know how we will survive without your wisdom, your creativity, and your political skill. But somehow we will make it, and I am so, so sorry I have clearly not supported you enough."

We held each other for a long time. Then she stood up, walked back to the clinic, and apologized. But I still did not know if she would stay beyond the week she agreed to; and though I had told her we would survive without her, I actually doubted that that was the case.

Even in the turmoil of the moment, I understood that there were structural issues that were also making Hotlin's life harder. In Indonesia, as in most of the world, it is difficult for women to take on a leadership role. Though the position of women in Indonesia is surprisingly good overall, it is not yet common for women to be in charge. Hotlin and I managed the staff (composed nearly equally of women and men), met with government officials (mostly men), and held meetings with village leaders (all men). I didn't quite realize how unusual it was though until one community man told me, "Since ASRI is a women-led organization, *now* we know that truly women can do anything."

If a score of 1 is perfect balance, and 0 is males getting all the benefits, Indonesia scores a rather terrible 0.17 in politics and senior leadership positions (but note that the World Economic Forum found the United States scored even worse at 0.16 in 2015!). This partly explains why people considered ASRI to be so unusual.

But while in most Indonesian cultures there tends to be a clear division of labor between women's work and men's work, women's activities are culturally valued. In Bali, for example, women do most of the religious activities that are considered foundational to the family's and community's well-being. In Java, women's property and custody rights are protected by customary law, and in places in Sumatra (the island where Hotlin comes from) some cultures are even matriarchal. In Kalimantan, women are in charge of all things domestic—often including the money. But interestingly, there appeared to be almost no conflict around gender roles in Kalimantan. In our initial rounds of radical listening, while women usually did not speak first, they always did speak, and men listened to their opinions attentively.

In Indonesia, there is also near perfect parity in education with girls and boys equally likely to enroll in grade school, and girls are even slightly likelier to start middle school, though there is a somewhat higher dropout rate for girls—often to care for sick family members. In later years, girls may also leave school because they are getting married—which unfortunately still happens quite young in some cases, even though child marriage is technically illegal. Our twenty-four-year-old gardener, Eka, wanted to marry a not-yet-fourteen-year-old girl. I was horrified, and I voiced my strong objection over many months— which, surprisingly, they seemed to be listening to. Finally, Eka's mother came to see me and explained, "Look, you have to stop opposing this marriage. Her family can no longer afford to feed her. Eka has a well-paid job with you, and he can buy food for both of them. She doesn't have an education, and she has no work. What is she supposed to do? What is her family supposed to do? And besides, I think they've been fooling around, and she won't be able to marry anyone else. They also love each other." What could I say? I agreed to stop objecting, on the condition that she would use birth control until she was sixteen (which she did, and then promptly got pregnant at sixteen, because she was bored and wanted to have something to do with her days).

Given this background of young marriages, it was particularly unusual for Hotlin to not be married in her thirties. Children are also beloved in Indonesia, and it is considered a great tragedy not to have them. This was a personal sadness for Hotlin, and she hoped very much to marry. A typical order of conversational questions when a woman meets anyone for the first time in Indonesia (perhaps on a bus, or even in the street) is the following: 1) "Are you married?" If the answer to the first question is "Yes" (as in my case), then the next question will be 2) "How many children do you have?" If the answer is "None," there is a pause. And then . . . 3) "Are you using *birth control*?" While I appreciated the openness, needless to say, answering these questions can be awkward if your life situation is not what you want it to be, as was the case for both Hotlin and me. Hotlin was looking for the right partner, and I was hoping for children (and not using birth control). Even the honorific you are given in everyday conversation signifies whether or not you have kids—Mother, Father, Auntie, and so on. So I could easily understand that this was an ever-present issue for "Older Sister" Hotlin, although we could both avoid the problem by just being called "Doctor."

In Indonesia, once a child is born, women do most of the child-rearing, but men are definitely more involved than was my experience in America. In the evening, bicycling home from clinic, I would often see men walking around with their babies—male or female—and showing them off to the other men in the neighborhood, who would coo admiringly. Wil and Clara—our first two nurses, who were married to each other—had their first baby about eight months after starting work with us, and Clara stayed at home for the traditional forty days of rest. She had a daily massage during that time (the most genius cultural custom ever). She was not allowed to cook. Her job was just to feed the baby and recover.

After his paternity leave, Wil came in to work every day looking haggard; he told me that, because Clara had worked so hard carrying and then delivering the baby, it was now *his* job to get up in the night and take care of their baby girl as much as possible. Only if she needed to nurse would he wake Clara. After three months, Clara came back to work, and their helper (a young Indigenous woman from Clara's Dayak village) would bring Holly to the clinic for breastfeeding. The thing that struck me was that they didn't ask my permission. I would have given it, of course, but they just assumed that this was the way it would be done. Every day, when the baby arrived at the clinic, the male staff would compete just as much as the women to hold her. Finally, Clara would get to nurse the baby before sending her back home. However, having babies constantly brought to the clinic may have made it even harder on Hotlin (and on me).

In addition to concerns about getting married and having children, what bothered me as well as Hotlin was that the world placed the focus on me, ignoring the huge joint effort made by so many people. Clearly, this was based in racism. The concept of a white savior was completely wrong; we were working from the premise that all our fates are inter-twined and that the solutions could come only from the communities rather than from outside. We were all guardians of the forest together. Moreover, the work itself was being done mainly by Indonesians. We had been growing considerably and now had more than eighty Indonesian ASRI staff. In the United States, Health In Harmony had added two more staff members (and luckily, the fundraising and communications to spread the knowledge about our work had followed apace). The only other outsider besides myself working on this project in Indonesia was Toni, who worked for ASRI doing grant writing while living in Java. Too many nonprofits hired mostly foreigners, with Indonesians serving only in low-level positions. That is not what was happening at ASRI. My long-term goal was also that I would fully pass over leadership to an all-Indonesian staff but, for the moment, Hotlin and I continued to colead.

The trick would be getting the world to truly see all the Indonesian staff doing such amazing work and, in particular, to see Hotlin. This was an issue both within Indonesia and outside the country. In Indonesia, my identity as a *bouleh* (someone once told me this literally means "albino") affected many interactions in complex ways. I was generally approached with a mix of respect and anger and even fear. This was not surprising; I was coming into a cultural and historical context that had suffered for nearly three hundred years under colonialism. All colonialism is exploitation, but people in Indonesia often say that the Dutch were a worse fate than the British. There are stories about how the British would come to negotiate a deal for buying spices, would then load the cinnamon and cloves on their boats, point their guns at the people, and totally renege on paying. The Dutch East India Company, on the other hand, would arrive, call a meeting for negotiating prices, promptly kill everyone, steal the spices, and enslave the population. For almost two hundred years, the Dutch East India Company ruled what is now Indonesia with extreme cruelty, while paying 18 percent profit per year on its stock—all, of course, to its wealthy European investors. They also consciously made sure that very few Indonesians had access to education.

This was largely why the communities around Gunung Palung were so poor; for hundreds of years, they had had their resources stolen from them, first by the Dutch and then by the Japanese. During World War II, the Japanese massacred thousands of people in our area and cut down the many hundred-year-old rubber trees that Sukadana had been famous for, to burn to make salt. Clearly resources had to flow back to them, and the decision for how those resources were spent needed to be in the hands of the communities.

My *bouleh* status was a two-edged sword that Hotlin and the other staff tended to use strategically, either keeping me out of situations or sending me in in cases where they thought it might be an advantage. We often strategized beforehand about the use of how my white skin could help them—or where it would be a hindrance. But we couldn't always hide me, and over time, most people knew I was there, even if I wasn't visible.

Many people viewed me with suspicion. There is a phrase in Indonesia about unknown agendas that asks, "What shrimp is hiding behind that rock?" The fact that I had given up a wealthy life as a doctor and wasn't even taking a salary was incomprehensible to most people. Many people assumed I must be there to make money. Maybe I was looking for some secret gold mine. If not that, then I must be interested in converting people to Christianity, since their only other experience with white people was with missionaries. But over time, it became clear to everyone—even the most suspicious—that ASRI was about neither money nor proselytizing. People came to realize that we treated everyone equally in the extremely diverse religious and cultural milieu around Gunung Palung and that we wanted *everyone* to have a better life.

Part of showing this to the community was modeling it in our staff. Hotlin and I worked hard to hire people from nearly every local ethnic group and to create a culture of acceptance and respect. Indonesia has the highest cultural variability on the planet, with more than seven hundred languages. With this level of diversity, it's remarkable the country could even exist—and partially for this reason, the government worked hard to promote tolerance. In school, children are constantly reciting the five core principles of the Indonesian nation, called Pancasila. These principles are: 1) a belief in one God (only five religions are officially accepted); 2) a just and civilized society; 3) one nation with cultural diversity; 4) democracy; and 5) social justice. This government policy has worked fairly well, although there have still been waves of ethnic and religious conflict across Indonesia since the 1960s, and there were always elements that pushed in the direction of extremism. Despite the official Indonesian principles, it was still rare to have the level of mixing of religions and ethnic groups that we had among our staff. But with our modeling of tolerance and equality, it was delightful to see the love and friendship develop between people from groups that rarely interact in Indonesia. This love extended also to the volunteers, who came from all over the world and represented many different religions.

I still remember one party we had down on the beach as a send-off for a Canadian volunteer. When I looked around the circle of people singing together, I realized that they could not have been more religiously and ethnically mixed if they had tried. A Jewish volunteer and a headscarf-wearing Muslim woman were tucked tightly together and snuggling in the same chair, a Catholic Dayak man laughed and played the guitar with a Maduran staff member (there had been ethnic violence between those two groups during the time I've been going back and forth to Borneo), and a Chinese Indonesian doctor belted out a song with one of our local Melayu cleaning staff.

Something about my upbringing had instilled in me the belief that change was possible. The belief in the possibility of a positive future is what I wanted everyone to have. This sense is what allowed me to try so hard to improve the lives of the people and protect the forest of Gunung Palung National Park—I believed that change was achievable, despite all the obstacles. But I didn't want to impose change, I wanted to partner in it. A quote from Lilla Watson, a Murri (Indigenous Australian) visual artist, activist, and academic, encapsulated my beliefs: "If you have come here to help me, you are wasting your time. But if you have come because your liberation is bound up with mine, then let us work together." I knew this to be profoundly true on the deepest spiritual level, but the climate crisis was also making it painfully clear on a physical plane that all our well-beings are intertwined.

What I was trying to establish were bridges where resources could flow back to these communities—but only under their direction, invitation, and design. We were attempting an anti-colonial approach to development and conservation, but given how rare that was, people didn't always get it. Nor were we always getting it right; the process was one of constant learning and I had clearly gotten it wrong in some ways with Hotlin.

When I first came, most of the community and staff were utterly fatalistic and pessimistic. This was not surprising, given that their experiences in life were with corrupt, exploitative, and sometimes threatening political systems. Democracy only really began in 1998, with the fall of Indonesia's second dictator, Suharto. The idea that power could come from the people was still very new. But over time, I was noticing a distinct change in our staff; when a problem arose, more and more they had faith we could find a way around the challenge.

I needed to learn how to use my privilege to focus more global attention on the impressive work of the Indonesian staff, including Hotlin. And I wanted the work to be recognized *within Indonesia* as being done by Indonesians—since that was the truth.

I also hoped that getting her more recognition would encourage Hotlin to stay—although, if she decided to leave to focus on getting married, I would support her in that. Getting another well-educated Indonesian to replace her would be extremely difficult. Hotlin's skill in government meetings, coupled with her devotion to ethical principles, were a wonderful combination. I knew that she loved ASRI despite all the challenges. She enjoyed making a difference in the world and working toward a society where people's lives were better. She was a gift to the program, and I hoped that ASRI would also be a gift to her.

16

A Much-Needed Retreat

H otlin did decide to stay—at least for a while. Her anger, though, made me realize that if I was going to hold on to staff, and keep my own sanity, I would need to slow down the pace, make sure everyone's contributions were honored, and try to give the staff more emotional support. I didn't want to see good staff quitting because they were upset. I had never seen a work space that centered on mutual respect and support, but I now focused my attention on making sure all the staff felt loved and on creating a healthier culture at work. I was managing to do a pretty good job with the team, but my own life felt like it was slowly breaking apart at the seams. I worried about my own capacity to keep going—and what that would mean for the continuation of the program. I tended to value the needs of ASRI and the community over my own and felt guilt and shame whenever I did eke out more time for myself.

I was at the end of my energy rope. I was making progress together with the team on community healing, but my base of personal healing was not strong enough. Hotlin seemed to be settling down and enjoying herself again after the blowup—but I was imploding. In July, Cam had successfully organized the huge Association for Tropical Biology and Conservation meeting in Bali, but things did not change when it was over. As just one example, our planned vacation never materialized. Instead, Cam chose to accept yet another invitation to speak at a scientific meeting. I began to feel that I simply had to organize a good chunk of time off, and with the encouragement of the Health In Harmony board, I decided to plan a two-month retreat. When I asked Cam if he would like to join me, he simply scoffed, and I realized I was relieved.

It didn't help that life in Sukadana, at the best of times, is difficult. The electricity went off at least five times a week, often for days at a time; the food is basic and unvarying; and the constantly invading creatures were exhausting. ("Ants in your pants" is real.) The village also gets up at 4:00 a.m. with the mosque, so there is never any sleeping in. Swimming was our only entertainment option; we had few friends who shared our mother tongue and our culture, apart from the short-term volunteers. The heat could be oppressive, and every single task—from purchasing necessities for the program, to processing required paperwork—took ten times as much energy as it would take almost anywhere else in the world. As a tiny example, I had replaced the clinic door handles every single year with the most expensive ones I could find, but they still all quickly broke.

While I loved both my work and the program, I was extremely unhappy in many other ways. The option of a retreat was only possible because I had been able to delegate increasingly over the years, gradually teaching the team to take on many of my tasks. One of our staff members, Ibu Patma, once told me that she loved having me as a boss, because I would assign her a task that was beyond what she thought she could do. She would work on it and then come back to me, and I would suggest changes and have her go and work on it again. After a few times, we would eventually get a good product. She said to me, "I know you could do this faster yourself, but I am so grateful that you teach us in this way, because every time you trust us to take on a project, we learn a new skill—and soon you don't have to help us at all." I was pleased that she saw it that way, because that was exactly what I was trying to do.

And our team members were also teaching themselves. We had hired a young woman, Ema, to be our cashier a few years previously who was unusual in having a high school education. She was our neighbor, living in the shack across the road. Of course, she had never used a computer in her life, but she was extremely smart and learned quickly. Per our protocol, she would enter every receipt—down to twenty cents—into our open-source accounting program, but then I still had to sort them into grants, a task that required speaking English and Indonesian, as well as understanding every grant and restricted donation. I also randomly pulled receipts and had staff go to the stores to check them to make sure there was no "leakage." I couldn't seem to figure out a way for someone else besides me to do these things, although I often worked late into the night on Saturdays to get through the accounting.

But then a remarkable thing happened. The moment the internet was installed, Ema began sneaking in at night to do an online university course in accounting (when Hotlin and I found out about this, we, of course, gave her permission). I began to wonder, though, how she might feel about training to be our accountant. It would take learning some English and more software. I felt real trepidation in asking her, because I personally hated the job—and I didn't want to ask someone to do something I disliked. But Ema's response was perfect and so unexpected.

"Oh, Dr. Kinari! That is exactly what I have been hoping for!"

And, indeed, she was soon doing an amazing job and loving it as I trained her in each part of the process. I was so proud of this young woman. One of our grantees even asked her to train their other grantees when they saw how meticulously she kept all the accounts, and our first auditor ended up donating when he realized how perfectly the accounting had been done. She lived with her widowed mother—and *ten* other family members—in the tiny little hut across the street from the clinic (although, with her increased wages, she began to expand it). Working with her taught me an important lesson about the benefits of delegating more of my responsibilities; just because I disliked it didn't mean someone else would. So thanks to the amazing team, and with Hotlin willing to take on full leadership, I was now feeling safe enough to try taking some time away.

I CHOSE TO DO MY retreat in Northern California because this was where I had done my residency, and I had a community of support there. One of my medical colleagues (and an original board member of Health In Harmony), Ann Lockhart, offered the use of her rustic turn-of-the-century one-room cabin, about an hour north of San Francisco. Also during my residency, I had found a Mennonite church in San Francisco that had become my spiritual home. This was a close-knit group of progressives who worked on LGBTQ rights and many other social justice issues. The congregation offered to support me through this time. I learned what that meant when I was picked up at the airport and provided with a bed and good food. And then, the next morning, I burst into tears and laughter when I was handed an envelope of cash—along with the keys to a cute white convertible, whose trunk was filled with groceries. Cruising north to the cabin with the top down, I felt one of those moments of stepping into a parallel universe, where everything you have given up is miraculously given back to you—and is even more appreciated for being so unexpected.

The redwood cabin, built in the early 1900s, felt to me like sacred space. Although I could walk to the library and grocery store, there were no other buildings visible from this grove. It was surrounded by redwood fairy rings and an untamed garden that merged into national forest. The fairy rings occur where a giant redwood had been cut (probably when the cabin was built), and then a ring of sprouts grow from the stump. Each of these sprouts can become an enormous tree in their own right, forming a magical circle of trees.

Visits by friends at critical moments kept me sane and surrounded with love. But most of the time, I was alone, and I threw myself into personal and spiritual growth with as much energy as I had given to ASRI. I read five or six library books all at the same time, went on long walks, and read my old journals. I ate mindfully, trying to be aware of every bite. And I had wonderful counseling; Peter Mayland (the retired psychiatrist who had helped Dardi heal from his eight years in a dark room) offered to see me without charge. I would go visit him, and we would somehow have therapy sessions that lasted seven hours at a time. He said he had never treated anyone this way before, but I seemed to be ready for it. I also had healing visits by my pastor, and I made a clarifying, but strained, trip to visit my mother and stepfather in Mexico. Toward the end, I spent three days in total silence at a monastery high on the cliffs of Big Sur. I don't recommend this kind of intensive retreat for everyone, but it was a profound experience of transformation for me. Buddhist monks take three months off every year, and I think there is great truth in recognizing the need to spend time reflecting inward, calming the soul, and seeking that still, small voice that always knows how best to guide us. I had truly been losing my balance, as well as my faith that everything would work out. My sense of being guided that had first formed during my year of inward reflection in the rainforest now, in the silence of the cabin, began to return with grace and strength.

And, as in the forest, I opened the lid and looked at some of the hardest stuff in my soul, this time going deeper. Most of my life, I had struggled with a profound sense of not being a good person—of being irredeemably flawed. I believed this in my inmost core, although I could see the disconnect with my equally deep-seated belief in the goodness of others. Part of me didn't really want to change, though, because I worried that this absurd belief might be part of what drove me to try to make a difference in the world. But the pain was at times intense, and I was concerned that my occasional lack of compassion for others began with a lack of compassion for myself. I knew people who knew me well saw me differently, but I simply couldn't take this in; other people's view of me did not ring true with my own understanding of self.

A book by Ken Wilber, called *No Boundary*, was very helpful in this struggle to overcome my conception of myself as unworthy. He says that whenever we set up mental constructs of dividing lines, they become battle lines. For example: Conscious/Subconscious; Mind/Body; Humans/Environment; and Divine/Non-Divine. But in fact, there are no strict boundaries, even down to the molecular level. Not even between energy and matter, or time and space. As humans, we love to make divisions, to categorize everything; but when we do that, we can actually create more problems than we solve. I was becoming more and more certain that separating out artificial constructs, like "health," "poverty," "psychosocial well-being," and "environmental conservation," had played a role in getting us into so much trouble on the planet. It truly is all *one*, and removing these divisions allows for holistic solutions.

Now I realized that I had to take those spiritual truths about boundaries into my internal landscape; I had set up divisions and battle lines in my soul that were not serving me well. I began to explore family constellations and to consider just how much our behavior, beliefs, and patterns come down through the generations, both in our learned reactions and genetically. I researched my own family, spending hours talking to my grandmother, parents, and aunts and uncles, and in the process, I started to see some of the same patterns that I had recognized in myself, even going many generations back: parentified children, the inability to see a child for who they were instead of who the parent wanted them to be, and a deep sense of shame. That process, in itself, helped me have compassion for myself.

But despite seeing its roots, I feared that the belief that I was a deeply bad person made me sometimes hide from others—not wanting them to know me too well, since if they did, they would surely reject me.

Maybe that also made me occasionally lash out, especially if someone was affirming what I most feared learning about myself. Cam was the normal recipient of these painful moments.

As I found the bravery to look deeper, I confronted some of the things I least wanted to look at. Memories had arisen in the forest, but I had quickly brushed them aside. Now I suspected that from the shadows of my subconscious, they might be wreaking havoc in my life. No one likes to talk about these things or even admit that they exist, but I also knew I was not alone.

I had been inappropriately groped when I was ten by one of the adults I most trusted in life and then raped when I was thirteen by another man when camping with a large group. The morning after I just remembered tears falling and my cold feet in cold boots cracking the crust of ice that had formed on the soft soil overnight. Was the "normalcy" of rape in my community the reason I shared with so few and those I did simply nodded? But now it no longer felt normal—it just felt so sad.

Of course, the men who did these things did not know, or probably care, that their actions would create incredibly long-lasting wounds. While the true extent is only beginning to be known, at least one in every four women are sexually assaulted during their lifetimes. I was surprised, too, to learn that being raped is even likelier to cause post-traumatic stress than the highest levels of combat. I suspected some of my belief that I was fundamentally bad came out of these experiences. I had never really gotten rid of that sick sense of dread, horror, and shame from these events. These experiences had also clearly influenced my relationships with men and probably were partly to blame for the intensity of my desire not to be controlled by them and paradoxically also being willing to put up with being deeply unhappy.

But like all things in our subconscious shadows, when we let the light in, they begin to lose some of their power over us and growth is possible. Similarly, the trees in our reforestation site flourished in the full light, but trees in the dark of the forest grew only extremely slowly. Blaming myself—and not even really knowing what I was blaming myself for—was just putting a sheen of ice over my life and not letting me thrive.

I spent three culminating days of silence at a Camaldolese hermitage on the cliffs of Big Sur, overlooking the stormy sea. I rose at 4:30 every morning for vigils, prayed while the monks chanted, and then wandered the steep scrublands, looking both outward and inward. Each day seemed to stretch longer than any day I had ever experienced, and the three days felt like an eternity. On the third day, after lauds, I sat alone in a circular wooden chapel, where light streamed down through a skylight, illuminating a suspended cross. I sat cross-legged on a slightly raised ringed platform that hugged the walls of the chapel. I was feeling bathed in all the painful things I had done and that had been done to me. Finally, I got the courage to ask the hardest question I could ask. Nearly shaking with trepidation, I managed to stammer, "God, am I a bad person?" I don't think I expected any response, but what I got was instantaneous, strong, and perfect. The Divine laughed at me! A deep belly laugh of joy and incredulity, forgiveness and love. It was so powerful that I began to laugh at myself, shaking as the whole space seemed to reverberate with this love. It felt as if an enormous, fossilized egg had been encasing my being, and this cosmic laugh cracked it open at last. I was hatched.

I teetered, vulnerable, back out into the world. In letting go of that belief about myself, I now would have to accept responsibility for living my life to its fullest, with compassion and love and laughter and freedom. I wouldn't be able to hide in shame; I would have to be willing to risk intimacy and the melted softness of life. I would also have to be willing to accept my full self and know that all of it was loved.

In the last days of my retreat as I started to forgive myself, I accepted one truth I had been struggling with throughout the retreat: I could not go on with my marriage as it was. While Cam and I still loved each other, daily life together was mostly miserable for me. He was disengaged from me and from his own soul. He worked insanely, traveled more than half the time, and the last years while he had prepared for the conference had been ones where I had been right—I simply hadn't had a husband. But things had not improved afterward, and we had ended up fighting most of the time.

In addition, I found myself increasingly frustrated with the structure of Health In Harmony. I was the head of ASRI, but at Health In Harmony, Brita was the executive director and my role was very unclear. In both my marriage and my work, I knew that the problem was not just the other. I needed to take more responsibility for things not going well. I could not ask *them* to change, I had to be the one to change—including bringing about a new structure if that was needed. Basically, I needed two divorces—with Cam and with Brita. For Health In Harmony, this work would be fairly straightforward, although painful. In long discussions with Brita, she, too, knew it wasn't working for her or for the organization, and she decided to move on to other work. I would have to officially lead Health In Harmony, which would align my decision-making capacity with my very real responsibilities.

But my marriage was a totally different thing—the prospect of giving it up was terrifying.

My insightful friend Rhiya Trivedi, who had been a volunteer at ASRI, liked to joke that I was really a nun. I knew what she meant, even though I was married with an active sex life. That year in the rainforest, as I began to examine some of the decisions I had made in my life, I became determined to give up the craziness of my youth and choose a moral path that focused on the well-being of others. But maybe I had gone a bit overboard—denying my own wants and needs to the point of just serving others—and sometimes resenting it. I appreciated the skills that having too much responsibility as a child brought to my life, and I didn't want to lose the compassion or the drive, but I did want how I cared for others to be a conscious choice, and I needed to learn to modulate it.

Even if my marriage was often deeply unsatisfying for me, the stability and care had been healing in many ways. Leaving my husband would also mean stepping away from my best friend—even if it had been a long time since that was a healthy friendship. I would also lose my beloved parents-in-law. Cam's mother, Jackie, was incredibly dear to me, and Cam's parents' farm in Sussex had come to feel like my own ancestral home. Even the work we did, risking so much financially, was partially rooted in the confidence provided by Cam's father, Stuart, who generously assured us that he had our backs. The idea of leaving that safety net was terrifying.

But sometimes, your heart will not be silenced no matter how hard you try to make it shut up and behave. The spiritual truth I had to learn was how to love myself, how to value my own needs as well as others'. But what if that path led away from staying in a marriage I was utterly miserable in? Rhiya had asked me a question that wouldn't leave me alone: "What if the best thing for you was also the best thing for others?" Yet how could it be the right thing to leave my husband? There had been times in our life when he had told me that he would commit suicide if I ever left him. Luckily, he had stopped doing that, but the fear for me was still there. Sometimes it felt like a cage, not a marriage.

The hardest part for me was how intensely he resisted nearly every step of the way at ASRI. Creating the program was hard enough, but fighting him as well made it so much more difficult. After our big fight about the organic garden, he had grudgingly allowed me to make decisions about ASRI, but he still made his opinions clear, and they were usually in opposition to whatever I was planning to do. I felt that some of the issues I had in our relationship were based on fixed personality traits that it wouldn't be fair to ask him to change. But could something be right for my partner, when it was clearly not right for me? Sometimes our deeply rooted ways of being just have to be dismantled—with as much compassion and love as possible—to make room for healthier patterns to emerge.

After much prayer and struggle brought me some certainty that God was actually leading me down this path, I made the choice. Then came the hard step of actually telling Cam. I hadn't spoken with him in weeks. He had responded to my taking space with absolute fury—but when I finally told him I wanted to meet him in person in Vermont and that I had arranged for loving friends' support for both of us—he got it. When he walked into the pastor's office where we would meet, I was surprised to see a man that I felt I hadn't seen since before he finished his Ph.D.: open, loving, and present. But still I knew I had to tell him my truth, and in the kindest way I could, I told my husband, after fifteen years of marriage, that I could not go on.

17

Royal Recognition

SUKADANA, INDONESIA, AND
LONDON, ENGLAND: MARCH–JULY 2011

Following our heart-wrenching separation, Cam spent the next month in the United States, while I stayed in Borneo. That month was a powerful time of change for Cam. When he arrived back in Indonesia, he said that my leaving was the best thing that ever happened to him. He felt like he had been asleep for years as he buried himself in work, and my leaving him had finally woken him up to what mattered in life. Indeed, it shifted his perspective more than I thought possible, and everyone who knew him was amazed at the transformation. "How can that possibly be *Cam* saying these things!" "He's usually so negative, but now he actually seems to believe change is possible and that humanity might have some value." Maybe people really can alter their basic outlook.

I began to feel that I was being unfair to him, and he begged me to give it another go, saying he was committed to doing everything he could to make our lives better. So I decided to try, and together, we began the real work of our marriage. We read the sort of self-help books

I never would have touched before, like *Men Are from Mars, Women Are from Venus* and *The 5 Love Languages*. I began to feel like a real idiot. You mean to tell me that some of our struggles were just standard things that many heterosexual couples grapple with? It was embarrassing to realize that if we had read these books earlier, they might have helped.

I was also finding myself profoundly changed. Continuing to look at repressed parts of myself was extremely painful, but bringing them to light helped me feel more grounded and whole. I was only just starting the process of spelunking in my subconscious, discovering a vast underground network of caves, and each integration of another part of myself brought healing to many aspects of my life.

This new internal strength helped me in my work. I had now taken on the role of executive director of Health In Harmony, in addition to leading ASRI with Hotlin's constant help. My goal was to bring efficiency to the Health In Harmony systems—increasing its capacity to share the vision, raise funds, and provide logistical support to ASRI. The work was necessary, but having two full-time jobs made it even harder to set boundaries around my work. This was partly because I loved what we were accomplishing and partly because of internal drivers that I wasn't yet willing to confront or able to control.

Hotlin, too, was starting to examine some of her inner demons. She began to realize how deeply she had been affected by the death of her brother in a bus accident when she was only twelve (he had been her closest confidant) and later by the death of her mother. She, too, had found ways to shield herself from others. Even though in many ways we were very close, it was only after nearly five years of working side by side that she told me about her brother's death. Hotlin was also dealing with issues of feeling valued and important in all aspects of her life and with issues of trust. Even though this was mainly an internal struggle, I knew that outside validation might also help—and that she deserved it. Toni and I had been working hard to try to get Hotlin that recognition. Toni found something that seemed just right: a conservation award from Britain's Whitley Fund for Nature.

IN MAY 2011, HOTLIN WENT to London to receive one of the most prestigious conservation awards in the world. The huge black-tie awards event was hosted by Her Royal Highness Princess Anne (Prince Charles's sister) and attended by David Attenborough and other dignitaries, to honor the seven Whitley Award winners from around the globe. After receiving her award from Princess Anne, Hotlin walked to the podium, dressed in a glittering traditional tailored long lace blouse (*kebaya*) and sarong. She took a moment to look around the room, taking in this surreal experience.

"As a dentist, I never imagined that I would receive a conservation award. As I am standing here, I remember a twelve-year-old girl who almost broke her jaw because of a chronic bone-related tooth infection. Her father brought her to our clinic a year ago for treatment, because he would not have to cut down the forest—home to the Borneo orangutan—to pay for his daughter's medical bills. Grateful for the recovery and in return for the care received, he brought us manure to plant trees in our reforestation projects. This project works.

"As I remember that little girl's beautiful smile, I remember that this community is our partner, along with my wonderful team in Indonesia and the U.S., to work together finding solutions to all root causes leading people to illegal logging. We believe this effort will be able to protect our beautiful forest in Indonesia. So in the future we can still enjoy, with a big smile, the wonderful forest where wild orangutans live happily, listen to the beautiful gibbon song, and with all your help, we can make it happen."

In her own remarks, Princess Anne highlighted Hotlin's work. "Perhaps the most important thing about today's winners is underlining the fact that you don't have to be scientifically qualified in biology or natural sciences to make a difference. Being a dentist is just as important!"

Returning to Indonesia, Hotlin became the star of three hour-long Indonesian TV specials on ASRI. She was featured in more than twenty print articles, including one in *Tempo* (Indonesia's *Time* magazine), and she received the national Kartini Award (named for an early Indonesian feminist). This public recognition was especially important because it relieved some of the pressure she was getting from her family. They now began to understand and value the work she was doing, even though she was not yet married.

The Whitley Award also provided a grant for ASRI that allowed us to follow the advice of the community leaders and hire staff in each *dusun* to help monitor illegal logging. We had started doing our own monitoring in August 2010 and had also found the perfect person to lead this effort: Bang Agus. He was a conscientious young man from Sukadana who had earned a forestry degree from the University in Pontianak. Agus cared about the forest, he was politically savvy, and he understood the ins and outs of community and government pressure.

Now we would be able to hire a local monitor in each *dusun* to work alongside Agus. In English, we would call these peer-to-peer change-assisters "Forest Guardians," though I prefer their Indonesian name, Sahabat Hutan (literally, "Best Friends of the Forest"). This new team of thirty people would work to help us find tailored solutions to the root causes behind logging.

One reason the village leaders recommended a local liaison in each village was because of the incredible diversity of the communities surrounding Gunung Palung. In our baseline survey, we found nine different sizable ethnic groups, plus many others with small populations. And each had its own culture and language. Using the national language, Indonesian, which is what I spoke, was sometimes of little use in trying to communicate. These Forest Guardians would be people who had grown up in their village and spoke the specific language of that village. This might make a huge difference in our capacity to reach people.

Moreover, these folks would know exactly who was logging and what the issues were in their lives.

To start the process of hiring them, Hotlin sent a message to the head of each *dusun* asking for five candidates. She requested that the candidates be selected on the basis of friendliness, honesty, love of the forest, respect within the community, and ability to read and write (at least at a basic level). Hotlin, Cam, and the conservation team interviewed each candidate and only a few times had to ask for additional options. Everyone was pleased with the final selection.

Cam, in his new mode of believing that change was possible, helped design the training, including guiding a trip to Cabang Panti so they could share appreciation of the diversity and interconnectedness of intact forest. The Whitley grant covered their salaries for a year, at thirty dollars a month each, for one-third time (quite a good salary, locally). After the first year, if the program seemed to be making a difference, we would have to look for other sources of funding.

Cam and the conservation team also designed a system to get more detail on what was happening with the logging on a finer scale than just *yes/no*. The Forest Guardians would keep track of which trails were actively being used to drag out logs, which collection points had wood stacked in them, how many people were cutting trees in the forest, and whether the sawmills were active. This data would help us assess whether logging was going up or down within the red villages.

The Forest Guardians would not be trying to impose change from the outside. We knew from the baseline survey that essentially everyone already wanted to protect the forest and the loggers wanted to quit—there were just barriers that were keeping them from doing that. We knew that our conservation education efforts shouldn't focus on trying to convince people about the importance of the forest but rather on helping people actually change their behavior.

I kept reminding myself of the stages of change that people need to move from: pre-contemplation to contemplation, then actually getting ready to try something new, and finally doing it (usually with some relapsing in the process). We hoped that the new Forest Guardians could help individuals move through each of these stages. They would serve as behavior change therapists providing support to the loggers.

My own recent work in struggling to shift long-standing patterns gave me even more compassion for how hard it must be for the communities around Gunung Palung. They were trying to move from traditional ways of farming and generations of being loggers to adopting new and unfamiliar activities. In my own experience, knowing for a long time that certain behaviors were destructive didn't make it any easier for me to change them. It requires enormous bravery, compassion for oneself, determination, and faith that there is actually a better way.

With the new Forest Guardians actively promoting our organic farming training, the demand for this training was increasing. The costs of the training sessions were covered by a foundation called Conservation, Food, and Health (when we saw this grant, Toni said to me, "If they don't *love* our program, they should change their name!"). Over the years, our organic farming training had become more important to people than we could ever have imagined. I saw this firsthand during the second round of Pak Ngalim's farming classes (about six months after the first round). After his opening talk, Pak Ngalim invited everyone to move into small groups to share experiences implementing what they had learned from the last training. The man sitting on my left—who had been looking quite pale—was just beginning to stand up when he wavered and started to collapse. I managed to catch him before he hit the ground, and as soon as my hands touched him, I knew he had a fever. We called the ambulance and brought Pak Din to the clinic. There, he proceeded to tell us his amazing story.

Twelve years previously, he had fallen from the top of a twenty-meter (six-story) coconut palm. The fall broke his back and smashed up his left foot. For two years, he couldn't get out of bed, but finally he "forced himself" to learn to walk again. His gait was still not normal, but he could get around. The problem was that injured foot never did heal; it had drained pus for twelve years! Every year or so, he would get deathly ill with fevers and shaking. This was what was happening to him now. The bacteria from the chronic infection in his foot bones was proliferating in his bloodstream, and if it wasn't treated, it might very well kill him.

One of the doctors put him on intravenous antibiotics and gave him lots of fluids, while our nurse skillfully worked on cleaning out the wound, cutting back the infected tissue and bone. The next morning, I came in to check on him and was delighted to see how much better he looked. "Good morning," I said, almost laughing with pleasure. "You look great!"

"Yes. Please take out my IV. I want to go back to the training right away! I don't want to miss anything."

"I'm sorry, but you don't look *that* good!" I told him with a real laugh. "You are still very sick. You need to stay in bed. Also, we need you to stay on intravenous antibiotics for a few more weeks, and when you go home, you will need to take antibiotics in pill form for even longer. I think we have a good chance to cure you, *but it could take a long time.* There is no need for you to go to the training. Just rest here in the clinic, and we can take care of you until you are well enough to go home."

"You don't understand!" Pak Din insisted. "In the first organic farming training, I learned how to make compost, and I doubled the amount I'm harvesting from my pineapples. I also grow herbs and eggplants, and now I don't have to buy expensive chemical fertilizers. I am making twice what I used to, and the cost of fertilizer is vastly cheaper than what I used to pay. I can even afford to send my children to school. *There is no way you are going to keep me in this bed.* I want to learn more!"

What could I say? After some negotiation, we came to an agreement. Pak Din would attend the next four days of the training, but he would sleep in the clinic every night and get antibiotic shots twice daily. The schedule that day included a visit to a successful farm owned by a previous trainee; I decided he and I would take the Kadie car so he could lie down instead of going by motorcycle.

The farm we visited stretched down a slope, with squash growing along the ridge, then rows of corn, followed by "long bean" vines supported on angled sticks (the beans are two feet long), then bright-purple slender eggplants, and finally, crinkly bitter gourds growing on bamboo trellises. At the bottom of the slope began the national park, with the forest stretching high into the sky and monkeys and birds calling from within. The group sat in a circle on the porch of the farm owner's little shack, and he told about his experiences after the first training. He used to spend about twenty dollars a month on fertilizer, but since he had learned to make his own, it was only costing him about twenty cents (one-hundredth of the original expense)—and his yields so far were actually *higher*. He now also understood how important the forest was for his water source, which was a well at the bottom of his property. He told us he forbade anyone to log near his land; I got the impression (although he didn't admit it directly) that he himself had recently stopped logging. He said that if the profits kept up, he would be able to buy a motorcycle so he could sell his vegetables in town and bring back manure from chicken farms.

Pak Ngalim pointed out that if they all pooled their money, they could rent a pickup and sell more vegetables and bring back more manure. Seeing their looks of skepticism, he made the whole group sit down and work out how much it would cost each of them per month. It was like watching light bulbs go on over everyone's heads. Pak Ngalim then suggested that they should rotate which crops each one was growing so they wouldn't flood the market with one type of vegetable. "This is why I've been telling you all that you need to form *cooperatives*. By working together, you will always be more successful than working alone."

The training group happened to include some farmers who belonged to a cooperative formed after the previous training, and Pak Ngalim asked them to share their experience with the rest of the group. The cooperative farmers talked about how they had converted a few old rice fields with badly compacted soil into successful vegetable gardens using raised beds. They were now working together to grow vegetables there, as well as farming their own land. They were sharing the profits from the joint farm, and a small portion went into their combined treasury. However, now the cooperative had a problem. They didn't have enough manure, although they were getting some from the local widows with goats. It would take them many years to buy a cow.

Ngalim suggested that he could help them write a proposal to the government for a cow. He had recently visited the department of agriculture, and they offered to help the farmers we were training—if the farmers could submit proposals. There were nods all around. Right there and then, they sketched out the key information for the proposal for one cow.

For several weeks afterward, Pak Din continued to come into the clinic for his antibiotic shots. It was such a pleasure to watch every part of his body slowly transform as the infection resolved; even his skin began to glow with a healthy color, and he stood taller and smiled more. At the same time, we got more good news. ASRI had received a request from the agriculture department to revise the proposal: they said we should make it for *ten* cows. Then, even more amazingly, they actually gave them twenty-one cows! This was more than the group could handle, so they passed ten of the cows to another village. Then the cooperative created a rotational plan to feed the cows, turn large amounts of manure into compost, and market their organic fertilizer, vegetables, and rice under a Green Diamond label. I was struck again by how much the lack of knowledge about sustainable agriculture and livelihood practices had been limiting these communities. But they understood what they wanted to learn, and as soon as they received the needed training, they jumped on it—and it changed everything.

It seemed that our integrated approach of tackling the problems from many different angles at the same time was bearing fruit, literally and figuratively. Pak Din's story illustrates this process perfectly. With the appropriate medical care—and a way to pay for it—*in combination* with the farming training, *everything* in his life could get better. To pay for his medicines, Pak Din worked in the clinic's organic farm and at the same time learned even more, like how to make natural pesticides from local plants. And now he could send his children to school, where they, too, were learning additional skills. Simply curing him physically would have been good. Giving him training in sustainable agriculture would have helped. Or just working with his community to protect the rainforest—the watershed for his rice fields—would have benefited him. But the combination of all three worked synergistically. Health In Harmony's board chair, Jo Whitehouse, once summed it up perfectly: "Synergistic means 1 + 1 = 11." Reversing the negative spiral requires both gaining new knowledge and an *internal willingness* to move to a new way of life.

On a community-wide scale, we wanted to improve the spiral of well-being for many, many people. We recognized that it was not just *what* we were doing that made a difference but also *how* we were doing it. We were building strong and respectful relationships, not only with the communities but also *within* the community (for example, through the farmers' cooperatives and the Forest Guardians program). At the same time, we were also helping to create a broader awareness that individuals could shape their own futures, as they saw one person after another do it successfully. When a community has that belief, the sky is the limit.

But to continue that work, we needed folks around the world to send their appreciation to these communities. It truly was a global partnership. The overall budget in Borneo had more than doubled over the past four years, from $150,000 to $320,000, but was still relatively inexpensive given what we were achieving (although I was still working for free). ASRI's income from patients had also increased from $28,000 to $72,000, but we needed the rest of the costs to be met by grants and individual donors. We were thrilled when we received twelve of the thirteen grants Toni had applied for—gratitude and support pouring in from around the world. In addition, we had funding come from generous individual donors.

In return for this "thanks" from global citizens, the communities around Gunung Palung were paying back large amounts of carbon by protecting rainforests and thereby leading to a healthier planet for us all. In truth, value was flowing in both directions, and for the first time since the program began, things were looking good. The staff were loving their work, and the community seemed delighted and empowered. Hotlin had received lots of recognition, and Cam and I were happier than we had ever been together. Even Dardi was finally doing well. He had gotten a job working in a coconut plantation and was sending money home regularly to his mother and sister, who had opened a little roadside kiosk and even renovated their home. His brain tuberculosis care, like all the things we had done in the program, had been a huge joint effort, and now the dividends were really paying off. I wanted to just rest and enjoy our success.

Around this time, Cam paid a visit to Borobudur temple on one of his trips to Java, and he texted me to tell me that he had just found out what my Sanskrit name, *Kinari*, actually meant. Apparently, it did not (as I had thought) mean "she who dances for the Gods"; rather, the word refers to mythical winged women who are *guardians of the tree of life*. My parents had chosen this name for me because they had met a woman named Kinari in Gujarat, India, shortly before I was born, and they liked the way it sounded. How eerie, and odd, to have all my names be so appropriate! I was asked to do an important task, to help guard the tree of life; but, like the disciple Thomas, I constantly doubted the reality of deeper layers of spirituality. Also like Thomas, I was granted experiences that confirmed for me that this was my calling. And then, once my name was changed to Webb when I got married, I began to work closely with others to create a *network* for guarding the magnificent trees of interconnected life.

Finding out what my first name meant just added to the nagging sense that the Divine was asking me to open up and listen even more. This had started during my retreat, a sense that there was something bigger being asked of me, behind these personal changes and the community work I was doing. I kept feeling called to work on a global scale. If this model worked so well here, it might work in other places as well. But I didn't want to do it. "Go find someone else!" I kept saying. "There are millions of more skilled people out there—go ask them! Leave me alone! You asked me to come here, and things are finally working. I'm living in a little village in Borneo because of you! What more do you want from me?"

The answer, I knew, was that I now needed to let others carry this work around Gunung Palung forward and begin to address the health of our earth on a much broader level. The tree of life, is, after all, not just one rainforest. Most doctors cure patients, a few also look at community well-being, but there were way too few planet healers. But I couldn't face it. I couldn't look at the horror of an unlivable planet for all humanity. That task was too big, and changing course was likely impossible anyway. In addition, I didn't think the team was ready to run everything without me. I was still seeing a false separation of "me," not the interconnected oneness of all. For six months, my ego steadily refused to hear this message. Some part of me knew I should listen, but my back remained resolutely turned.

And then, as we often did, Cam and I went down to our little cove in the evening to enjoy a lovely swim in the ocean. But this time, God's electric touch was a little more literal.

18

Hope

TEGALALANG, BALI:

APRIL 2012

L ying in bed in a one-room bamboo house in Bali, I propped myself on enough pillows to do the tiny bit of work I could manage each day. Beyond my feet, I could see Cam typing away at a table and then a beautiful view of terraced rice fields.

It was already ten months since my jellyfish sting, but I was still having trouble facing my limits and setting boundaries for myself. I was proving to be a very slow learner and a poor patient. I understood clearly that pushing myself too far, either mentally or physically, could mean days of recovery, but I still found it hard to limit myself to just one or two hours of "work" per day. My world had shrunk to the confines of one bed in a tiny bamboo hut where I could hobble about for short periods of time and a lovely pool where I could get a bit of exercise each day. (Horizontally, my heart could move the blood around my body, but when I was upright, my blood vessels were not properly constricting.)

After being hospitalized twice in the United States, I had made it back to Borneo—barely. I abandoned the second leg of my flight on to Indonesia and had to go straight to the emergency room in Singapore.

After five days of rest in a hotel, I somehow managed the short flight to Borneo. From my bed in Sukadana, I began to transfer all my responsibilities.

It soon became clear that I could not stay on in Sukadana. Though I was no longer in pain, merely standing would bring my pulse to 150—a level indicating extreme exertion. Walking would drop my blood pressure so low that I had to squat or even lie down on the spot. Even so, I always felt guilty if I did not go to our clinic in the middle of the night when a patient was severely ill, though I knew that the Indonesian doctors could handle it well and that our medical visitors were on hand for consultation. If the patient died, I would worry that maybe I could have done something. So I would try—but I was soon so run-down that I ended up being hospitalized in Singapore. That was when Cam and I decided that we would have to relocate to Bali for me to be able to fully rest and hopefully recover. My friends in America were strongly encouraging me to come back home, but I knew that my body could not handle another twenty-five- to thirty-hour transpacific flight.

I felt grateful for Cam's constant care, and I knew how difficult it was for him to do essentially everything in our lives. Without his help, I very likely would have died. He had been by my side constantly for most of those ten months (after his heroic trip to Boston) and was now delaying a two-month field trip to the island of Seram because I wasn't yet able even to cook for myself. We had been, unfortunately, unable to find anyone who was willing to come help take care of me so he could go. Cam's mother, Jackie, was now rearranging her life, though, to come to Indonesia in a few months so that Cam would be able to do the travel required for his grant.

In January, when I was still severely ill in Boston, the board under my medical school friend Alison Norris's chairmanship had decided to hire an executive director and found Michelle Bussard. She was already doing a fabulous job managing the staff that I had increased to five, and it was a relief to leave things in her capable hands. Likewise, at ASRI, Hotlin had stepped up to the leadership role (with hands-on, capacity-building support from Michelle, in several trips to Sukadana). Hotlin and the rest of the team were stretching themselves to cover everything I used to handle—hiring and firing staff, overseeing the finances, making sure activities were on schedule for various grants, and envisioning where to go next. I was extremely pleased that, with minimal input from me, everything continued to function.

My help was still needed in some things, though. My main task, that morning, was to review Thomas Lazzarini's preliminary results from ASRI's five-year survey. Thomas had just graduated from Yale when he was recruited as a volunteer to help with the survey—thanks to a chance encounter with board member Peter Mayland asking for directions in one of the Yale quads. It was an enormous task, conducting a comparison survey to our baseline from 2007. I had planned to guide Thomas through every step of the process, but now he was largely left on his own with just a few easy-to-say-but-hard-to-do bits of advice from me: "Go hire at least twenty student nurses from the nursing school, train them in survey and interview techniques, randomly sample households in every village, and then organize transport and accommodations to all thirty villages. Good luck. I have to go back to bed." Truly, it was almost that sketchy! But Etty Rahmawati and Dr. Ruth became Thomas's support system, and the three of them pulled it off superbly—training the nurses, coordinating everything, and getting interviews in 1,497 households. Thomas was now in the final, critical stage of the project: analyzing the data.

Opening Thomas's files, I looked at his summary of key findings. Reviewing the first section, my eyes kept getting wider. He *had* to be doing something wrong. I barely looked at the rest of the numbers and fired off an email:

There can't possibly be a 68% decline in illegal logging households in just five years! I do think it's gone down, but that much? That would be a decline from about 1350 logging households down to 450. I'm sorry to say, but it must be an error. And the health data are improved across the board. How can that be? You also found that loggers were significantly likelier to have quit if they knew about the ASRI program. I want that to be true, but is it? Can you send Cam and me a file with how you are coding everything? Let's try to work this out long distance, but you may have to come to Bali.

Lying there and thinking it over, I realized that, at least for the logging, I might be able to get some corroborative data from other studies to help me figure out whether these findings could possibly be right. As I opened up each study and reviewed the data, I was surprised to see that Thomas's analysis might actually be correct. An Earth Observatory of Singapore survey, in late 2011, reported that 2.4 percent of households were logging—a rate nearly identical to (and even slightly lower than) Thomas's survey results in early 2012, showing a 2.9 percent rate. That suggested that we were not seeing a bias in people's responses to interviewers associated with ASRI. When I compared these more recent rates with a published study by Marc Hiller in 1991 and some internal surveys by Esme Cullen in 2008, I saw a nice downward curve over time—both in active logging rates and in households that had ever participated in logging. I sent the comparative data off to Thomas and tried not to hope too much that the data was correct.

I WAS HAVING TROUBLE STANDING up, but there was no way I was going to miss our anniversary party, celebrating the opening of the clinic five years ago. I was also especially excited because we had confirmed with Thomas that all the analyses were right. Cam and I had flown in the day before, and I had rested as much as I could. The piece of land where we planned to build a medical center now hosted a big tent for the party, erected over the site where our demonstration garden used to be. The same land also held ASRI's conservation office, where the staff had thoughtfully made a spot where I could lie down. I wanted to be outside, greeting the attendees, but if I didn't rest now, I would never make it through the whole event.

About a hundred invitations had gone out, but we expected many more people to show up. By the time we were ready to start, at least three hundred people had already gathered, including representatives from the regency government, the Department of Health, the national park office, all our farmer groups, the village heads, the community health workers, and friends and neighbors. Our head nurse, Pak Wil, helped me teeter across the site, picking my way through the soft earth in my high heels and a formal silk sarong and *kebaya*. My opening speech was followed by a few words from the *bupati* (the head of the regency), which was read by his representative. He expressed how grateful he was to be partnering with ASRI, and I noticed he used the word *ethical* a number of times. Hotlin and I smiled at each other, recognizing this indirect reference to the fact that we never paid bribes. We were pleased that he knew this and that he actually appreciated it. Hotlin had even developed a clever technique of asking for a donation anytime a government employee asked for a bribe and assuring them that we could promise them that every penny would go to helping people and that there would always be a receipt for every single transaction. They usually immediately backpedaled.

Looking out at the crowd, I also got lots of smiles from government officials, many of whom were our patients. Each year, there had been more and more collaboration between ASRI and the government. We had done three trainings of government midwives and one with the traditional birth attendants. After the trainings, whenever I met any of the midwives, either in town or out in the villages, they would stop me to say how grateful they were and tell me about when they had used their new skills—a number had also joined the event today. The health department, also well represented, was providing us with free birth control supplies, immunizations, tuberculosis medicines, malaria treatment, and some lab supplies.

Now, sitting in the tent, we listened to a few speeches, including one from the head of our local *desa,* who had been active in helping us design the program. And then a giggling group of children came up to sing—the first cohort in a brand-new program we called ASRI Kids.

The previous summer, a doctor from Stanford, Dr. Ewen Wang, visited ASRI along with her two children, Ana Sofia (age thirteen) and Lucia (age nine). The two girls were willing to help out in small ways—but coming to Borneo had not been *their* idea, and they weren't too happy about it. A few days after their arrival, I saw them sitting together on the porch of a house we rented next to the clinic. They were playing the card game Uno, while a group of local kids stood around them, looking over the girls' shoulders and trying to figure out what they were doing.

"Why don't you teach them how to play?" I suggested.

"But we don't speak Indonesian!" Lucia protested.

"I bet you can do it with pointing. Give it a try!"

Sure enough, when I passed by an hour later, ten kids were sitting in a big circle, laughing and playing Uno. That was the beginning of six weeks of powerful bonding among them all. Ana Sofia and Lucia learned how to climb coconut trees, find stick insects, and catch cicadas. The local kids experienced doing handstands, playing games, and singing new songs. By the end of six weeks, they were inseparable.

Nearing the end of their stay, Ewen invited a few ASRI staff to come along with her family on a memorable visit to the orangutan rehabilitation center in Central Kalimantan. The whole group loved the adventure, but Ana Sofia and Lucia were dismayed to see that the only Indonesian tourists in Tanjung Puting National Park were the ASRI staff they had brought with them. When they got back and showed pictures to their Indonesian friends, they were even more dismayed; with Etty translating, they learned that most of the children had never seen an orangutan even though they lived within just a few miles of them. The kids also had no idea that orangutans are found only on the islands of Sumatra and Borneo.

After they left Sukadana, Ana Sofia emailed me to present an idea.

Suppose they came back the following summer and organized a summer class for the local kids? They would learn about the rainforest and how to recycle trash, and also some general health and hygiene. Then, at the end of the summer, the group could visit the orangutan rehabilitation center. I loved the idea, and I hated to have to tell Ana Sofia that I didn't actually have funds for it. Not to worry, she assured me; *they* would raise the money. And they did! The girls threw fundraising parties, and Ana Sofia sold cards of her gorgeous photos from Borneo.

This summer class was instantly a huge success. The kids loved it so much that they showed up for class an hour ahead of time every day. Etty did a brilliant job, teaching the curriculum that she and Ana Sofia had designed together. And the trip to Tanjung Puting National Park was clearly the highlight. My favorite photo shows the look of open-mouthed amazement and sheer joy when one of the kids first caught sight of an orangutan. I remembered that feeling myself.

Cam also took the class on a hike to the top of the mountain that rises behind the ASRI clinic to show them the biggest and most magnificent tree in the area. As they hiked, they learned all about the gibbons, hornbills, many-colored insects, and fascinating plants around them. But when they finally reached the tree, all that was left of it was a stump. The stump itself was so big that all the children were able to climb up on it together—some of them with tears in their eyes. (We later found out that it was probably the father of one of the children in the class who felled that tree.) The children were so upset that they went the following day to talk with the newspaper, which made a front-page story of the children's opposition to the logging.

Now the twenty-five children in that class were standing at the front of the tent, wearing their bright green ASRI Kids T-shirts—proudly decorated with their own drawings about the natural world. Their first song was one written by Lucia, in English and Indonesian: "ASRI Kids, oh, ASRI Kids! Look at us, we've got the future in our hands, with our ideas we'll take a stand!" I glanced over to see Ewen watching with moist eyes as Lucia led the song. The next song was a popular Indonesian song with the refrain "It's all gonna be okay!" Finally, all the kids, joined by the staff in the audience, sang the Alam Sehat Lestari song. That song was written as a gift to me by one of our doctors, Dr. Made, and it never failed to make me tear up, with its lines about the beauty of the natural world, about living in harmony with it, and about ASRI helping to make that happen.

I looked across at our multiethnic, multireligious team, laughing and singing. Sitting among all these people, I realized that, even if our program ended tomorrow, we had made some changes in the communities that, with luck, would last the lifetimes of the people here. And for the children singing, I hoped their lifetimes would be long.

Next up onstage was Etty to host a few talk show–style interviews with people involved in various aspects of our work. Her first guest was a widow who had received her two goats four years previously—and who now had thirteen! Then came a patient who had been paralyzed by tuberculosis in his spine and was now proudly walking again. Following him, a farmer spoke about how he had done a carefully controlled study with one plot of rice grown organically and one with chemical fertilizers: the yield had been higher with the chemicals, but his profit was double using organic methods due to lower costs and a higher sale price. Finally, Etty interviewed the head of one of the villages, Pak Bastarin. His responses were so fascinating and were corroborating the survey results so much that I pulled out my notebook to take notes.

"Five years ago," he began, "when I helped design this program, there were more than a hundred loggers in my village, and now there are fewer than ten. Loggers can now become farmers without start-up money, because they can make fertilizer from the waste around them. All along the road, where there was just grassland five years ago, there are now organic vegetable gardens and rubber trees."

I knew Pak Bastarin was right about the farms along that road, because I had traveled it many times on the way to our reforestation site; indeed, farms were sprouting up everywhere. And Thomas's analyses—that we had now confirmed—showed that over half of the ex-loggers were now farming. But it hadn't occurred to me that the lack of start-up money needed to buy fertilizers could prevent loggers from switching to farming. Pak Bastarin's insight confirmed once again that the communities knew exactly what the issues were and how best to solve them. In those initial radical listening meetings, they were already thinking many steps ahead to issues that I was just beginning to understand five years later.

From the survey, it also looked like the shift away from logging wasn't making people poorer. Daily income was now about $1.47. We didn't have a good measure of daily income in our baseline survey; the best comparison data we had was from a 2003 regional WHO survey, which reported $0.46 a day—so this was a dramatic increase (though still well below the absolute poverty line of $1.90 a day). Another indicator that people were indeed better off was the overall increase in ownership of consumer goods between our two surveys; cell phone ownership went from 20 percent to 80 percent, and motorcycle ownership went from about 45 percent to 85 percent, transforming households' ability to participate in the local economy and get to our clinic. I'm sure that ASRI helped with this economic transformation, but we weren't the only factor, with the government working hard and overall wealth increasing in Indonesia. Still, many people had stopped logging without getting poorer.

Following Etty's talk show, awards were given out to three Green Knights—local residents who were organizing their communities around sustainable solutions. These heroes, greeted with strong cheers, were hugely proud of their plaques. Hearing the enthusiasm from the crowd, I was reminded how critical these celebrations were. They provided perspective and helped us all see the progress that was being made, and they honored all the effort that so many people were giving to bring about change.

The Forest Guardians in their ASRI batik uniforms were also proudly standing about. They were doing a great job, although there was still work to be done in making sure that everyone in the communities knew about the discounts for non-logging villages, since only 22 percent of respondents were aware of the red-green system (as worded in the survey). Given the decline in logging, this could mean that the discount wasn't necessary and that just providing access to affordable care and training in alternative livelihoods was all that was needed. Or it was possible that the logging discounts did matter and if more people knew about it, we would have seen an even rapider decline in logging. But among those who clearly knew about it, more than half had discussed the discount with someone in their village, and 96 percent of those believed that the discount had helped stop logging in their village. This is exactly what we had hoped: that the discount would help build social pressure to encourage loggers to switch to other occupations. I had also been pleased to see that now 93 percent of people thought the park was a good thing (up from 79 percent).

After some closing remarks and songs, we invited everyone to have lunch: rice, spicy tempeh, green beans, a chili sambal, and a small piece of curried chicken served in a folded banana leaf. Dr. Nur went to get one for each of us and sat down beside me to eat. Nur had graduated at the top of his class from the best medical school in Indonesia, and he was one of the smartest human beings I had ever met. He had agreed to stay at ASRI for another two years in return for a four-month opportunity—specially tailored for him—to study at Yale and Stanford. The plan was to give Dr. Nur more training so that he could take primary responsibility for teaching—not only ASRI's new crop of Indonesian doctors but also the Yale medical students and resident doctors who were doing six-week stints of learning at ASRI. Naturally, Nur was feeling nervous as well as extremely excited. Could he survive without eating rice three meals a day? Would he be able to handle the snow in Connecticut? And was his English good enough? Smiling at him and remembering my own fears about coming all the way to Borneo, I assured Nur that he was going to be fine.

From the ASRI Kids, to the Green Knights, to Dr. Nur's training in America, the day was a perfect illustration of how all the parts of the program were interconnected, not only health care and livelihoods train-ing but also educational opportunities, both for the communities and for our staff. I think one reason things were working so well was that we had managed to create a supportive environment—both within the community and for ASRI's staff and volunteers—that empowered people to manifest their own greatness. This was as true for the loggers as it was for professionals like Dr. Nur and even for visiting children.

I could see this sense of empowerment reflected in an evaluation survey conducted by a conservation volunteer named Erica Pohnan (who came to us with a master's degree in forestry from Yale, followed by a year of forestry education in a top Indonesian university). Most of the people she interviewed believed that ASRI was the key factor in stopping logging in their villages. One man summed it up by saying, "After ASRI came, we began to have hope."

This sense of hope may have also come because people were gen-erally healthier and fewer were dying. The survey found that ASRI pa-tients were half as likely to be worried about accessing and affording care, and people were overall healthier. Nur asked me about the results of the survey, and we began discussing the key points. He was delighted to hear that the percentage of households with infant deaths declined from 3.4 to 1.1—a huge drop in just five years. (And sure enough, I'd felt a bit vindicated when I had shown Cam in Bali that births per mother had also significantly declined from 3.1 to 2.6.)

Nur and I talked about how the drop in infant deaths was proba-bly due to better curative and preventative care but also to improved economic well-being and possibly even to a healthier natural environ-ment. The survey showed that more than two-thirds of children were now immunized. This was mostly thanks to the government, but Nur thought that ASRI could take some credit for their improvement, too. The head of the regional department of health once told Nur that, until he saw the ASRI clinic, he didn't know it was possible to provide high-quality health care in a remote place. "I wasn't even trying," he said. And one of the doctors at a government clinic whom I had spoken to before lunch had told me that he motivated the staff by telling them, "Look how good ASRI is. We can be that great, too!"

Nur asked me about the results of the survey questions about symptoms. We had asked about household disease symptoms in the prior three months, including fever, diarrhea, unintentional weight loss, or a cough lasting more than three weeks. While none of these symptoms map directly onto a specific illness, fever gives us a hint at malaria, diarrhea indicates water quality and hygiene, and the last two were often signs of tuberculosis.

"You won't believe it. Fever went down 71 percent!" I told Nur. "We've both seen how rare malaria cases have become in the clinic, especially since we distributed the mosquito nets in 2009. I wonder if it has anything to do with the logging dropping, too?"

"Kinari, I went through the database, and we haven't seen a single local case in two years."

We then talked about how diarrhea rates went from 40 percent of households to 20 percent. There were a number of potential reasons for this, since nearly everyone was boiling water now, up from only about a third when we started, and many fewer people were defecating in the river. Nur speculated about whether digging wells in many villages had helped and, again, whether a healthier forest led to cleaner water.

"But what I'm really curious about is the tuberculosis symptoms." Nur leaned toward me. Tuberculosis is just about one of the worst diseases one can get. Seeing people waste away to skin and bones while constantly coughing is heartbreaking for any doctor and, of course, they become a huge economic burden on their family—not to mention the risk of infecting everyone around them.

"It's good. Very good. In 2007, about a third of households said someone in the house had been coughing for more than three weeks within the prior three months. Of course, that could be many things, but it gives us a hint at TB, especially when combined with unintentional weight loss. Now only 10 percent of households said they had long-term cough, and weight loss also dropped from 19 percent to 6 percent of households. I know it's weird I remember all these numbers, but I've been thinking about them so much."

Nur proceeded to tell me that patients diagnosed in government clinics would now also be seen by our community health workers to make sure that they took their medication routinely. The government was excited to work with us since we had excellent success getting people to take their medicine for the whole course. In return, the government would provide all the medication for free, including for our patients.

Nur laughed. "This is extra good for me because Yale is making me take preventative tuberculosis medicines because my test showed that, not surprisingly, I've been exposed." Hearing Nur say this reminded me how important our work is globally. If even one patient gets on an international flight with drug-resistant tuberculosis or another infectious disease, they can spread it all over the world. The well-being of everyone on earth really is intertwined.

At that moment, though, my own well-being was starting to flag, so I began saying my goodbyes. I was escorted to the ambulance to go home and lie down. I was feeling very light-headed and suspected my blood pressure was low; I took my pulse and found my heart racing into the 180s. It was working flat out just to get enough blood to my brain, even though all I had been doing was sitting in a chair.

From my bed that afternoon, I thought about the day and the future, wondering again about the tugging I kept feeling to work on a broader level. And it did seem that I was needed less and less. Each time a key staff member moved on, I had been terrified that we wouldn't be able to continue. And each time, equally amazing people showed up to fill the gap. Our new dentist, Dr. Monica Nirmala, was a perfect example. After only a few months, it was clear that she was going to be an outstanding leader. She had come to ASRI because, like Hotlin, she wanted to do even more in the world than care for people's teeth. A friend once said to me that replicating our approach might only be possible if serendipity were also replicable. I was beginning to wonder if it was.

As long as we continued to work with passion toward a better world, remarkable people just seemed to keep being drawn to that mission.

Over the next few days, I would be helping the staff make their three-year plans. The staff had been conducting evaluation meetings with their constituents and would now make their own goals. It was already very clear that ASRI Kids should be expanded. Improving education had been among the community's requests in our first radical listening meetings—and now, Etty was hoping to add seven schools to the program each year. Over the next three years, nearly all the twenty-two schools surrounding the national park could be reached. Ana Sofia and Lucia pledged to continue to raise money for ASRI Kids, and they hoped to return next summer.

Most of the communities seemed to be very happy with the ASRI program, but we did hear some grumbling, mostly from villages where illegal logging had dropped dramatically but was not yet at zero. I could understand people's frustration; if their village went from over one hundred loggers to fewer than ten (as in Pak Bastarin's village), it still did not qualify for the green discount. That was why some people were asking for a third status between red and green—say, yellow—which would qualify for a partial discount. Specific criteria for the yellow status would also have to be discussed. We all knew that listening could never be "one and done"; every part of the program could continually be improved.

It was kind of like listening to my body; once was not enough. And after the long day of celebration, my body was telling me very clearly that an extended rest would be necessary—possibly weeks in bed. Hopefully, I would be able to marshal enough energy to help the team plan their quarterly goals. But if I couldn't do it, I was now learning, they could probably pull it off on their own.

19

A Prophetic Dream

Back in Bali, my health deteriorated, and it was very unclear how long it would take me to recover or if I ever would. Facing mortality yet again was forcing me to be as honest as possible with myself and try to learn to listen more to my limits. It was strange to have expectations for my own life shrinking while my understanding of possibilities for the earth were expanding.

I recalled a conversation I once had with a new volunteer. We were traveling out to a village, and as we rounded a curve in the road, we were greeted by an exquisite view of the forest stretching up into the mountains of the park. The volunteer asked me if I really thought Gunung Palung National Park could be saved—and even if it could, would it really matter anyway? Wasn't this all hopeless? Why was I even trying?

My answer at the time was quite pessimistic. "I will be very surprised if we can decrease the illegal logging. And if we can, and if we do manage to save some of the biodiversity, honestly, I believe that it will be for some post-human world. It looks very likely to me that humans will destroy themselves. Population growth, ecological collapse, increasing divisions between rich and poor, and an unwillingness to recognize what's happening—it's the perfect storm." I took a deep breath and looked at her. "But the only thing in my hands is *this* day, *this* life saved, *this* tree saved. I believe that we still have to give our whole lives to doing the right thing, in whatever way we can. We may be smashing just one small hole in a huge wall. But someday—like the Berlin Wall—maybe enough of us will break down parts of our false sense of separation from the natural world, and we will be free. All we can do is try, even if it is hopeless."

Most days, I believed my prediction of doom; I saw humans as inhabiting a kind of petri dish, in which we would continually expand our use of resources until they ran out. Those were the beliefs I lived with daily (and which Cam reinforced). But recently, a feeling of dangerous hope had been seeping in. I say *dangerous,* because, if there was some real chance of averting disaster, I knew I would have to face up to that quiet, persistent voice that told me that transformation might be possible on a vastly bigger scale and that I should start working on that broader level.

As humans, we struggle to grasp how radically things can change; as soon as one problem is solved, we focus on the next. I knew this in my own soul. I tend not to look back, to remember where we started from, or to give myself credit for all the changes in my own life. I could only see the next issue to work on. We forget to recognize how far we have come: from a world essentially continually at war, to the greatest period of peace the world has ever known; billions have been lifted out of extreme poverty; life expectancy keeps rising; and women now can vote in every country in the world (New Zealand was the first in 1893 and Saudi Arabia the last in 2015). Obviously, we still have much yet to do, but we should also see where we have come from (Hans Rosling has fabulous interactive graphs that illustrate all these points). However, much of the progress of humanity has come at the expense of the natural world and only because of the use of fossil fuels.

The problem is, in most rights revolutions, a core group of disenfranchised people usually begin by putting themselves on the line. But the birds, insects, orangutans, orchids, and rivers can't argue on the streets or in the courts that their well-being should be considered. Even so, many humans are stepping forward to advocate for them and for the health of our planet. I personally know many, many people who are passionately committed to defending the natural world and even more who are its silent supporters.

Nor is this true only among the wealthy of the world. I will always remember one woman who came into our clinic from a nearby village; she puzzled our registration manager when she didn't want to be seen as a patient but asked to see me—and refused to say why. When I sat down with her, she started by saying how much she appreciated that the clinic cared about the natural world as well as saving people's lives. Then she made me promise that if she told me something, I would make sure that the right people got the information. I agreed. With some trepidation, she told me that someone she knew had killed an orangutan, and that the male's head was still in his house. By this time, a few other staff members had gathered around to hear her story.

Some men she knew in her village had been hired by a businessman in the capital city of Pontianak to hunt an orangutan. Apparently, the trader planned to sell the meat at very high prices to rich people who wanted to eat exotic bushmeat. The hunters went into the forest and searched for days until they found a big male orangutan in a tree. She told us that she had heard how the first shot only wounded him and that, before they managed to kill him with another bullet, the orangutan used some leaves to try to stop the bleeding. "He was just like a human! We use leaves from the forest to cure our ailments as well. He was using traditional medicine! And after they brought his body to the village, I saw his hands. They were just like human hands. I am so angry at them for killing the orangutan. How could they do that? I don't want to turn them in, but it isn't right. What they have done is just wrong." As she spoke, tears ran down her face.

The staff and I were crying as well. Many of us were thinking about Riki, the baby orangutan whom we had treated in our clinic. International Animal Rescue, another nonprofit near us that cares for orangutans who were losing their habitat to the palm oil plantations encroaching south of the park, brought him in while their veterinarian was traveling. Riki had been rescued after his mother was shot trying to escape from the bulldozers. He was severely dehydrated, and our nurses had no trouble finding a vein for an IV (the veins were in exactly the same place as in a human baby). It appeared that he may have also been shot, because I felt small, hard objects in his belly. We started him on antibiotics after consulting with the local Indonesian veterinarian, and he perked up enough to be taken back to the recovery center. Sadly, though, Riki had not survived, and an autopsy indeed showed bullets in his abdomen. Our tears were not only for these two but all the thousands of orangutans dying each year in Borneo.

After sharing her story, the woman proceeded to give us exact directions to find the remains of the male orangutan. That information (along with other details gathered by another nonprofit called Yayasan Palung) was critical in making a case against the bushmeat trafficker, who was finally arrested with untold blood on his hands. Catching this culprit had a widespread impact, since these traffickers are actually rare. That woman's bravery, along with a dedicated community, helped stop a poaching ring that operated far beyond Gunung Palung.

There will always be some people who are willing to do almost anything for money, whether out of desperation or greed. But once social norms change enough, the public no longer tolerates activities that were once accepted. When we first started our work around Gunung Palung, most of the communities already wanted to save rainforests—even though logging was still unchecked. I'm not sure where these beliefs came from, but they were quite strong. In our survey, 92 percent of non-loggers were opposed to cutting down trees, and amazingly, two-thirds of the loggers agreed. That meant that the large majority of people who were logging didn't even consider it a legitimate activity.

They also nearly all said they would prefer alternative work. This is notable because there is considerable variation in the ethnic groups surrounding the park with successive waves of migration into the area. Yet there wasn't any variation by religion or race in their desire to still have thriving rainforests near them. If one defines Indigenous people as those who believe they belong to the land and not the other way around, all these groups would likely fit in that category. This worldview likely has very deep roots in all the cultures in Indonesia but is also reinforced by living daily in proximity to the natural world and being dependent on it. However, there had also been about ten years of educational and conservation outreach work done by the nonprofit Yayasan Palung, who had helped break the poaching ring. Their work may have helped tilt the scales before we began our radical listening sessions. (Yayasan Palung was started by another orangutan researcher, Cheryl Knott, who has continued to conduct research through Harvard and Boston universities.)

But even though these communities wanted to protect the forest, that wasn't enough; on their own, they didn't have some of the knowledge or the resources to bring about positive change within changing economic realities. In theory, these problems could have been solved by the government (especially if they knew what the community-determined solutions were). Our local regency had over the years been increasing access to medical care and even providing free services to many patients, but still there often ended up being unexpected expenses, especially if a patient was referred to another regency. I wondered if Indonesia carried through on its plan to provide universal health insurance by 2022, whether that might have a big impact on illegal logging throughout the country. They planned to have citizens pay monthly premiums, but maybe someday we could convince them to allow people to be able to pay their premiums with seedlings. It would even be possible in an integrated government system for forest communities to get discounts for their premiums based on logging rates. In my perfect world, these discounts and the seedlings would be paid for by carbon offsets from companies and governments around the world that cared about a sustainable future for us all. Indonesia would get help with health care, and the world would get a healthier planet.

Something similar could work in the developed world, too, where taxes for universal health care could be prorated based on carbon footprint. This would make visible the true link between human and environmental health and how we all have an obligation to both.

Would we realize that we simply *had* to learn to live in balance with our earth or face an un-survivable future? Would our love and compassion for the natural world triumph over our desire to look the other way? My reading and reflections during this forced rest were starting to shake my belief that this change was impossible.

Maybe Indonesia could lead the way in showing the world how to do it. I lay in bed with the possibilities buzzing in my brain.

SINCE I JUST WASN'T GETTING any better, I started seeing an acupuncturist. On my first visit, she told me, "Kinari, the universe went out of its way to give you this sting. It was a *gift*. There is something you are supposed to do differently. A great change is being asked of you. Probably God tried to get your attention in other ways, but because you weren't listening, more had to be done to get you to change."

This seemed profoundly right, and yet, I could still feel myself resisting. In the second year of being primarily bedbound (although improving with the acupuncture), something happened that pushed me even harder: I had a dream. I awoke deeply disturbed, remembering every tiny part of it in perfect detail. And unlike normal dreams, this one did not fade one iota afterward.

In the dream, the entire world was represented by a large valley where people had gathered around lakes; some lakes had small fish, and others had huge ones. But all the lakes were being poisoned by runoff coming down from the surrounding mountains, which were under government control. Soon there would be no fish at all. In the dream, I was leading a small band of rebels, mounted on horses and carrying muskets, but we were no match for the military with its fighter planes, troops, and tanks. With many of us injured, we had to escape to the foothills to recover.

Evoking the New Mexico landscape of my childhood, these cliffs were filled with the ruins of ancient Indigenous peoples. I knew I had to hide in a cave to gather my strength. Others were already trying to fight back from their cave hideouts, but I realized that uncoordinated resistance would never work. Isolated in unconnected caves, we were too easy to attack. But surely, my dream-self reflected, our ancestors knew this as well; they must have developed some "back passageways" allowing them to work together. I just needed to follow these paths of wisdom that would provide the solutions for how we all needed to co-ordinate. Creeping out from my cave, I found a secret staircase carved into the cliff; a Native American woman standing at its base greeted me warmly and allowed me to go up the staircase. Ascending the stairs on this ancient path, I immediately became overwhelmed with doubt. It was all hopeless anyway. And who was I to be here?

It was then, in my dream, that I felt the voice of the Divine. Crystal clear. "If you do not choose to do this work, the human species will not survive. You must choose."

In the dream, I struggled with doubt and fear. "Me? How can any one of us be that important? How can any of us matter?" But the message would not let me go: "You matter. Each of us matters. We are only part of the collective whole, yet our choices matter." As much as I struggled and refused, my dream-self knew I must do this. It felt like a very long time of resistance in the dream but finally, trembling, I said yes to the calling. Slowly, a feeling of acceptance and determination seeped into me.

I found the back passageways. I sent children as messengers to the valley. "Collect as many people as you can!" I told them. As folks arrived, I sent them down these passageways to help coordinate our activities so that the military could not predict where the resistance would come from next. We won the battle.

But the war was not over. In the next scene of the dream, cities all over the world were resisting, each with coordinated activities. Our group had gathered in one of the cities, and we held the skyscrapers. Troops in riot gear were massed in the streets. "How can we win?" someone asked as she looked down in fear.

"We cannot lose," I answered. "We hold the high ground." From that metaphorical high ground, we dropped water balloons on the troops, who were forced to go around the white walls of truth we erected in the streets.

Eventually, we won. In the last scene of my dream, I was walking across a desolate plain, and I met a resistance fighter burying his father. I mourned with him at the graveside and told him how deeply sorry I was to have led them into this. But he told me that his father was grateful to die for something he believed in, so he was proud to bury his father in this way. Then, going farther across the plain, I met a government soldier who was bleeding badly. I lifted him to try to get him to a field hospital, but he was clearly dying, and I set him down. He looked at me with love and said, "I am so glad we lost. I was fighting for the wrong thing. I am ready to go."

Following this vivid dream, I went through an intense period of fear that persisted for many weeks. Did I really have to make a choice? Did I have to be willing to call people together to work toward a healthier way? Me? Who was I? This was really absurd grandiosity. I wanted to just say, "No!" And anyway, it was only a dream and didn't make logical sense, and I was sick and weak, recovering in my cave. But I was unable to shake the sense that this was important and that I would regret it if I said no.

I decided to do a day of fasting and silence.

During that day, what came to me was that this was not just about my choice but about the millions of people that needed to hear and join together. What I chose to do did matter, and so did the choices of so many others. I emerged from that day with what felt like a download about a declaration of a commitment to the earth, a declaration for myself and others to sign. It reminded me of how the Founding Fathers of the United States signed an unprecedented Declaration of Independence—but now, the work needed to be global and no one could be left out. A Declaration of Interbeing (to borrow a term Charles Eisenstein uses from Thich Nhat Hanh) for our planet.

But that day of silence didn't quiet my fears. I alternated between concern that I was completely delusional and feeling crushed by an overwhelming task. I didn't feel like I could sign what I had been asked to sign. When I told my acupuncturist the dream and the message about the declaration, she suggested she take me to consult Ida Resi, a young, low-caste Balinese woman who had recently become a high priestess—even though, in Bali, only high-status, wealthy, male elders can become high priests. The Divine had opened for her a firmly closed door.

I had nothing to lose. Maybe Ida Resi could help me interpret the dream and its message. After a long car ride, where I lay down in the back seat, we arrived in the remote village on the slopes of Mount Batur with stone-walled compounds surrounding extended family units. We were soon sitting before an unassuming twenty-three-year-old woman dressed in a traditional Balinese sarong and lace *kebaya*. Her hair was done up in a swirl with a jeweled spike in it, and she wore a necklace of jade, which I later found out came from the Dalai Lama. We talked for a short while as a group, and then I asked if I could speak to her alone. Sitting before her on her ornately carved porch, I began by telling her I had had an encounter with the Divine many years ago in the rainforest of Borneo. My plan was to then tell her a little bit about my work and finally ask for her advice about the dream. But before very many words left my mouth, she held up her hand with palm toward me and stopped me.

"I have a message for you."

She knew almost nothing about me except that I could speak Indonesian. She did not know I was a doctor, my dream, or the work of planetary health I was involved with. She composed herself to speak, looking directly at me with solemn, shining black eyes and the kindest smile.

Then her words flowed. "You have been chosen. You have been chosen to help heal the earth. You have a message you have to give to the world. Your heart already has much wisdom and strength, but your mind is stronger, and it is fighting with your heart. Your mind is blocking your acceptance and surrender. When your heart and your mind become one, and the wisdom flows from your mouth, then you will have pure peace."

I sat there with tears running down my face, unable to speak.

Ida Resi then told me that she well understood how hard it is to be chosen, and she assured me that the path would always be opened for those who answer a calling. While there would be many difficulties along the path, overall it would be more amazing and fulfilling, she said, than I could possibly imagine. However, if I continued to fight it, I would get even sicker. She told me the calling was a blessing and that I had to accept it. She told me I was afraid of what others would think (and she was right) but that I must stop thinking that way. I would have to find ways to be in alignment with my inner voice of wisdom in all aspects of my life. She asked me to meditate, and then she prayed over me for a long time, singing and chanting. And in that meditation, I could feel my "Kinari" wings expanding before all my boundaries dissolved, and I knew, for a time, the oneness of all.

20

Planting and Fire

2012–2013

From my bed in Bali, I was still helping the ASRI team think through challenges, but they were doing more and more on their own. And as my health slowly improved, I recognized that, even if I *could* eventually go back to managing ASRI full-time, I *shouldn't*. Founders play a key role, but as an organization grows, different skills are needed. Neither Hotlin nor I really found joy in managing a maturing organization. What we were good at and enjoyed was grasping new ideas and a big vision. It was time for us both to make room for people who delighted in making systems work efficiently and effectively.

One of these people was Dr. Monica Nirmala. Promoting ASRI's dentist to serve as executive director would free Hotlin to do fundraising, based in Jakarta, and would allow her to continue her near-yearly fundraising trips to the United States. And it was also clear that, with more than one hundred staff, ASRI would need a middle management team. We promoted our head nurse, Pak Wil, to be head of human resources and our DOTS coordinator, Ibu Lia, to manage logistics, drivers, cleaners, cooks, and maintenance staff. We also hired a volunteer, Erica

Pohnan, as conservation director with a two-year contract; she became ASRI's only non-Indonesian staff member, overseeing our organic farm trainings, forest monitoring, childhood education, and reforestation activities. In addition, she would be able to assist with some of the things that could be difficult for non-native English speakers to do: write grant reports in English, provide language and cultural translation for the Health In Harmony staff, and help navigate any challenges our Western visitors might have.

After my dream, I started tentatively encouraging the Health In Harmony and ASRI teams to research the possibility of doing similar programs in other seriously threatened ecosystems. While I was still scared of what this calling might mean, I realized that, regardless, the first step would need to be to prove whether the approach that had worked so brilliantly around Gunung Palung would do the same elsewhere. Did local communities everywhere know exactly what the solutions were to live more in balance with their environments, and would meeting those needs produce such dramatic results? I knew it wouldn't be hard to find places where the need for health care led to overexploitation of the environment since I had seen many sites like this in 2005 when I traveled across Indonesia. In fact, I had barely seen anywhere where this wasn't the case. Now we focused on identifying promising areas for a second site in Indonesia.

Health In Harmony and ASRI decided in May to send an evaluation team to Indonesian Papua to the group of islands called Raja Ampat, where they found the need for health care was a key driver of coral reef and forest destruction. While the team was in Raja Ampat, I continued to recover in Bali but because Cam needed to travel again, a Canadian friend, Rhiya Trivedi, came to stay with me. After volunteering at ASRI, she had finished college and then taken a year traveling around the world for a Watson fellowship, including to India where her parents grew up. It turned out to be mutual aid, as she was recovering from a traumatic experience. We had a profound time of emotional intimacy and psychological healing, which also made me realize how much that joint journey was lacking in my own marriage, even in its improved state. We then traveled to Sukadana, where I would support Dr. Monica in her new leadership role and where Cam would meet me when he returned.

The crisis happened almost immediately after Cam arrived. I was propped up at the head of the bed, with the roosters calling out the window and the draped-up mosquito net waving in the hot air from the fan. Cam came in, and after a kiss, he sat at the end of the bed and asked me how I was doing. I excitedly began to tell him all about the internal spiritual work I had been engaged in with Rhiya and about a transformational water blessing ceremony we had done with Ida Resi, the high priestess whom I had been seeing regularly. But he quickly shut me down, saying that he was sick of hearing about it after years of my explorations.

I hung my head, not knowing what to say.

Seeming to try to pacify me, he said, "What are you feeling about the calling to work on a broader level?"

I rallied and began to tell him some further thoughts, but within moments, he was arguing with me about how I should go about it.

"Cam, I can't do it anymore. I've told you this before. I just barely have the energy to do the work, but I cannot do it *and* fight you as well."

We sat in silence for a very long time, my misery clear to us both.

Staring intensely at me, he seemed to gather his courage. "Kinari, you have to be brave enough to tell me what you really want. You *have* to be that brave."

The unexpected question rung me like a bell, the sound reverberating down into my deepest core, cracking things open I didn't even know were there.

I slowly moved into a cross-legged position. Tears started slipping down my face, and then the floodgates opened and I began sobbing like a two-year-old. I had never cried so hard in my life. The kind of crying where my whole body racked with pain and snot dripped from my nose. Kindness and love flooded Cam's face, and he held my hand

and gave me a tissue. He waited, his gaze intent. I felt like there was something hidden in the depth of my solar plexus and that part of me was reaching into the darkness, trying to grab hold of a truth that was hidden deep within. When I pulled it up and saw what it was, fear filled me, and I resisted, trying with my whole being to shove it back down—but once seen, it would not be silenced. My voice cracked a few times, then, finally . . . "I just want to be alone."

Once the words were out, the tidal wave of crying faded, and calm overcame me. We stared at each other, stunned.

Cam's voice broke, and he half whispered, "I don't think you have ever said anything truer to me in your life."

Though I wanted to take it back, my mouth just opened and closed. I knew he was right. We both looked at each other with huge love, but this new reality was slipping between us, and the gulf was widening by the moment.

"I will move out."

I just nodded.

Then began forty-eight hours where we lovingly cared for each other, cried for hours, and talked about our life together and all the blessings it had brought us. We worried how I would survive on my own since I could still not care for myself either physically or financially. When I had taken a salary for the first time the year before from Health In Harmony, the five thousand dollars had all gone to health expenses. But there was no choice, and we both knew it. Among the mourning, there was also excitement. I knew there were important things I needed to explore about myself, spiritual paths I wanted to go down, and journeys my soul was calling me to take on my own. I also knew that this decision, even though painful, was in full alignment with my soul. Two days later, on our eighteenth wedding anniversary, Cam moved out.

About a year before, I had had a dream during a nap that I had not understood until this time. It was one of those dreams I immediately recognized as a dream I had had before. I was a red wolf—half coyote, half wolf—who had been the head of the pack until she fell in love with a human man. She loved him so much that she transformed herself into a beautiful woman. From that union flowed many children. But in the dream, I knew I was not truly a woman and that I could no longer keep myself from turning back into that red wolf. So with love for my husband, I transformed back. The children grew well knowing their parents' love was strong enough to turn her into a woman—but not forever—and now she raced wild and fiercely free. On moonlit nights, she would still howl to him from far away.

My union with Cam had produced many wonderful and strong figurative children—like the Health In Harmony and ASRI programs. But it was also true that I could no longer hold back the transmogrification into a truer self. The day he left, I danced around the house in joy and then collapsed on the floor and cried—mourning mostly what could have been but somehow never was. Now new possibilities begged to be known.

In the months following, with long-distance care from Cam and lots of friends' support, I was able to survive—and even thrive. When Cam left, I still couldn't stand up for more than twenty minutes, but shortly after our separation, I actually began to recover energy. Rhiya returned with me to Bali, and soon I could walk a few blocks, then start doing exercise. By the time she left a few weeks later, I could care for myself. I felt I was being shown the beginning of what was possible when my heart and my mind aligned and the message flowed from my mouth.

THREE MONTHS LATER, IN SEPTEMBER, my health was good enough that Hotlin and I decided to go for a five-day evaluation trip to Central Kalimantan to visit another organization that was asking for our expertise in assessing the needs of a series of villages along a river. The river formed a dividing line: the land to its east had been devastated, first by timber companies and then by palm oil plantations; but the west side was a critical peat swamp area—logged but salvageable.

The stories we heard on that trip were heartbreaking. One old man told us, "In the logging days, we were all so rich, but now all we have are memories." Another man told me that some people were earning up to two thousand dollars a month, during the peak of the logging—and then he talked about the night when he and his friends blew through more than a thousand dollars on drinking, karaoke, and prostitutes. No doubt few people realized how fleeting that wealth would be. Now, people said, they understood that the forest is what gave them true wealth. I wondered how much that was a metaphor for the world. We think the resources are endless and that the high life will continue—but without a healthy ecosystem, the roaring turn of the millennium may come crashing down faster than we could ever have imagined.

When the valuable timber was all taken out, the palm oil companies had moved in. They cleared what remained of the forest, usually through burning, and then drained the peat to plant palm oil trees. When peat swamps are drained, they begin to decompose, releasing enormous amounts of carbon. The drive for inexpensive cooking oil for unhealthy processed foods was making this process repeat across much of Borneo. Incredibly diverse forests were replaced by monoculture, which would be very hard to bring back to natural forest.

Now these communities were desperate to keep the forest that remained by partnering with outsiders to preserve it. It was only in the loss of half of their forest that they recognized the true value of what was left. The peat forest on the other side of the river held enormous stocks of carbon, large numbers of orangutans, and, for the local communities, their only chance at a promising future.

The palm oil plantations were also eroding people's capacity to live in their traditional ways; the fish were poisoned by runoff, and much of the rice field land had been bought by the plantations. People were often forced to take wage labor with the palm oil companies, but prices at the company store were so high that, along with other deductions, employees described often going home with nothing left from their $150 monthly checks or even with a negative balance.

Nor were these communities alone. Shortly before going on the trip, I had seen an article posted on my favorite environmental website, Mongabay, showing that pro-environmental views were in fact widely shared throughout Borneo. In a survey of people in 185 villages across Borneo, two-thirds didn't want deforestation because of the negative effects on their livelihoods and on the forest.[1]

A former research assistant at Cabang Panti told me about his own experience working as a manager for a palm oil plantation after he stopped working in the forest. He quit, he said, because you either yelled at people or you were yelled at. No amount of money was worth that kind of life. Instead, he became an organic farmer, using the skills he learned at ASRI. This was also the hope of these communities in Central Kalimantan: that with new skills, they might create independent livelihoods. With more economic freedom, they would have the choice to keep their forest.

In radical listening sessions with random groups of fifteen to fifty people in each village, we asked them what the solutions were to protect the forest. Only this time (and in Raja Ampat), we were very careful to avoid telling them we were health care providers to make sure the results weren't biased. We found that consensus was reached after about an hour and a half in each village and that, like at Gunung Palung, every village along the river reached the same conclusions. The three critical needs that emerged were: 1) more education for the existing health care providers; 2) fish-farming training; and 3) water filters. People felt that with these three things, they would be *mandiri*—independent or self-reliant. The word implies having the ability to control one's own fate. Access to these things, they said, would make their lives stable enough that they would not need to sell the remaining forest to the palm oil companies and would not have to "become slaves to them." They would be able to protect the forest for the world's well-being and their own.

---

1 See: Erik Meijaard et al., "People's Perceptions About the Importance of Forests on Borneo," *PLOS ONE*, September 9, 2013.

While the communities seemed to know exactly what the solutions were, they lacked the necessary knowledge or resources. However, unlike around Gunung Palung, these solutions would not require a long-term commitment on the ground. All they needed was to be provided with some training and to be connected with existing resources. They didn't need a new large-scale project like ASRI. Hotlin and I passed on the information we learned to the organization that invited us. In theory, they would be able to provide these needs on their own, but we kept the door open in case they ended up requiring further assistance. We knew that Health In Harmony would never be able to go everywhere in the world, so part of our goal was to demonstrate to other nonprofits how this work could be done and teach them to do it on their own.

In consultation with the Health In Harmony and ASRI boards, we decided not to work in Raja Ampat, mostly because in the triage of the planet, rainforests probably needed to come before coral reefs, since the greatest threat to coral reefs was actually ocean acidification. Saving rainforests was also more of our expertise. But it was sad to discover that the need for health care was a driver of destruction in ecosystems other than rainforests (one man in Raja Ampat described collecting more than 220 pounds of dried sea cucumbers to pay the two hundred dollars for transportation for a medical emergency, even though he knew that would decimate future harvests).

What we ideally needed was a way for communities all over the world to be able to share their solutions with nonprofits, governments, companies, and individuals who might be able to help them meet those needs. However, I knew that this would require a change in mindset. The organization that invited us asked Hotlin and me how we were going to assess the community's needs. When we told them we were going to ask, one of their staff members was flabbergasted: "But these are uneducated, poor people. They don't know anything. You can't ask them!"

It might help to convince them to trust the communities if we could show strong outcome data from other sites. But I also suspected that for many, this would not be enough. In most development organiza-

tions, it didn't take much digging to find attitudes of colonialism, racism, and patriarchy. I certainly hoped that the organization we had just
done the assessment with would actually implement the community's
solutions—but I wasn't entirely sure they would.

RETURNING TO BALI FROM THAT trip to Central Kalimantan, I stepped
onto the airport tarmac, looked up at the full moon, and took a deep
breath. I almost felt like howling. For five days, I had been sleeping on
floors, holding daily radical listening meetings, carefully balancing on the
sloping planks that formed bridges to the floating outhouses, and traveling at least six hours each day over bad roads or on the river. I might
need a week of recovery, but I was pleased to still be standing.

Looking up at that full moon, I was overwhelmed with thankful-
ness for my whole life. All the hard things and all the good things. I
had been letting go of layer after layer of attachments, and these lessons were helping me be more present and feel gratitude. Despite more
emotional lability than I was used to, I loved beginning to know truer
and deeper aspects of myself.

Walking across the tarmac, I turned on my phone and saw a message
to urgently call ASRI's conservation director, Erica Pohnan. I called her
right then.

"Kinari, I don't even know how to tell you this. There's been a fire in
Laman Satong—a terrible fire. The community tried to stop it like they
have all the others, but it was too big and they just couldn't. So much
work planting those trees! So much love. And it's almost all gone."

"Did anything survive?"

"Only about 10 percent—luckily, the oldest part—but everything
else. . . . it's just ashes." At Laman Satong along the southern boundary
of the park we had planted about sixty acres of denuded land over the
previous five years.

"How? How did it start?" I asked Erica.

"We don't know. It's possible it was just from a cigarette, although some of our staff suspect it may have been arson from a mentally ill man in town. Either the fire jumped the firebreak or someone walked across—we can't tell anymore. By the time the community got there, the flames were huge. They just couldn't control it. I'm so sorry." I could hear the tears in her voice.

Then Erica told me that when Hotlin had arrived that morning (she had an earlier flight and less distance to go), the two of them and Cam (who was in Sukadana packing up more of his stuff while I was away) had immediately driven the hour out to the community nearest the reforestation site.

"We sat in the Forest Guardians house, and people just kept coming in. Nearly everyone in that community has helped in some way with the reforestation, and it felt like they all wanted to be there. It was so sad. We all just cried together."

The effort to reforest that area hadn't even just come from that whole village. Many thousands of other people had contributed through growing the tens of thousands of seedlings—as payment for health care or in exchange for mosquito nets. Every year, we had hosted a big Green Day event at that reforestation site: poems were read celebrating the environment; children performed traditional Dayak dances, wearing hornbill feathers in their hair; government officials gave speeches recognizing ASRI's work—and then the hundreds of participants would all plant seedlings together before sharing a meal. One year, the former village chief and a published poet, Pak Johanes, read a poem describing logging as a wound to the earth and ASRI's work as helping to heal that bleeding tear. The bandage of all those planted trees was now gone. All that effort, year after year—tending to seedlings, preparing the ground, weeding and tracking their growth—had mostly gone up in smoke.

"Kinari, they all want to replant."

"What is the point?" I moaned. "It could just burn again. Do you have any idea how much that cost? How am I going to tell our donors?"

"Kinari, we do radical listening! This is what the community is begging for. We have to listen. It is important for all our healing—even if it is lost again to fire."

I sighed deeply. I knew she was right, and I was glad that she was living ASRI's values. We should do it if the community really wanted it and if our donors would support it, but it was painful and felt hopeless.

"You know I always quote Mahatma Gandhi that the ends are never in our hands, only the means. But boy, is it hard when the ends turn out so badly." I thanked her for telling me and promised we would talk again soon.

Then, moving into the shade under the awning outside baggage claim, I dialed Cam. I could hear the pain in his voice. "First you leave me, and now this! I feel like everything I care for most is being stripped away from me! Could the universe have designed this more perfectly?"

For me, too, this felt like another painful letting go. But the loss was harder for Cam. For him, the reforestation site was a child he had nurtured from infancy, giving it much more attention and care than I had. He had designed the trials, advised on the seedling composition, and spent hundreds of hours under the hot sun, helping each step of the way. We cried together, mourning what wasn't to be.

Then we talked about how, possibly—just as grace had come to me, in the throes of death—this terrible fire might bring some unexpected gift. Cam shared something one of the men in the community gathering had said. Before helping with the planting, this man always thought if they logged, it would be fine—because they could just replant. But now, he said he understood what a huge effort it was to reforest, how expensive it was, how slowly the seedlings grew, and that it could all be lost in a day. "He said that he would never cut another tree." Maybe this profound lesson could be learned no other way.

We consoled each other that at least the other reforestation site was doing well. And Cam was going to make sure that the firebreak at our other site, the "orangutan corridor," was wide enough. That site had once been the orangutans' "highway," connecting two areas of their habitat until a new series of slash-and-burn rice fields broke apart even that strip of forest. So with the communities' agreement, we had replanted those fallow rice fields with tree seedlings. Erica had always called this site "the good child," because the trees had grown back so much more quickly, likely because of the more closely surrounding forest and the wet, swampy soil. Amazingly, once the densely matted grass and ferns had been cleared out and seedlings planted, we began to see orangutans traveling through on the ground—captured on our camera traps. These automatic wildlife cameras even caught an orangutan doing a selfie as he checked out the camera and two sun bears' butts as they meandered past. We hadn't seen this level of wildlife in the Laman Satong site, but bird surveys showed an increase from three species to sixty, and there was clear indication that wild pigs had been entering the area. Even though it might take thousands of years to restore the amazing lowland rainforest, it seemed that usable *habitat* could be restored relatively quickly—if fire could be kept out.

Unlike northern forests, mature tropical forests don't naturally have forest fires. They only burn when they have been logged, leaving lots of deadwood, or when they are still regrowing. In the tropics, the longer you can prevent burning, the less likely it is to burn at all. In Laman Satong, we feared we had just been reset to zero, and fire would now be even likelier.

Cam and I agreed that we would talk again the next day. "Kinari, I am so glad you are still my friend, and I'm really looking forward to seeing you in Bali next week. And even given this fire, I want to thank you again for choosing Gunung Palung. We've made a huge difference here. I'm sorry I doubted you for so long."

"You're welcome, Cam. And thank you for saying it. I can't wait to see you next week, too."

"I should be mostly done packing up my stuff." I heard a choke in Cam's voice. "But is it okay if I leave most of my gear and plant specimens in the front office?"

I assured him that of course it was and that we would continue to share both the rented house in Bali and our Sukadana home. For now, he would be living back in Bogor, where he had friends who could support him.

"At least meditation is helping right now. I know you've been telling me about its benefits for years, but apparently, it took you leaving me for me to finally try it." He laughed at himself, and I smiled. Cam's anger at spirituality seemed to be softening at the edges, and the opening to new ways of being that started the first time I left had increased even more so with our final separation.

"It's just so much right now. I'm going to leave a book for you here; it's called *The More Beautiful World Our Hearts Know Is Possible*. I signed up for a workshop with the guy who wrote it, Charles Eisenstein, when I'm in Bali. I know you will like it."

Hearing this from Cam was disconcerting, but it made me deeply happy and gave me hope. I'm not sure why he had never been interested in this sort of thing while we had been together, but maybe it was true that the best thing for me in the end would be the best thing for him. The burning of our marriage was allowing for new growth in Cam—and in myself. We closed with an agreement about completing the paperwork we would need for finalizing our divorce in Guam.

That day was a roller coaster of emotions. Following the high of envisioning new growth, both personally and for Health In Harmony, came the pain of such a deep loss. Precious hopes and dreams had been invested—as seedlings, planted over years—and were gone overnight.

I grabbed my backpack off the carousel and met my regular taxi driver, Pak Nyoman. As the car wound its way up the lower slopes of the volcano, illuminated by the bright moon, we passed an impressive statue in a roundabout. This dramatic Hindu statue is one of my favorites, and in that moment, it perfectly captured my conflicted feelings.

A chariot is plunging down a slope, drawn by six terrified horses. The traces have come undone, and each row of three are veering in different directions. The charioteer has drawn his bow, aiming it at an enemy who stands perched defiantly on the heads of the first two horses, raising his magical dagger. This is the exact moment when uncertainty reigns: Will the horses come under control, or will the chariot be torn asunder? Will the hero shoot his arrow in time, or will he fall to the enemy's dagger?

I asked Pak Nyoman whether he saw the statue as depicting the conflict between good and evil. His response was perfect. "Oh no. We Hindus do not believe in good and evil. Both creative and destructive forces are necessary. Most things are gray, and it is hard to know in any given moment how things will turn out. That is what the statue is about. The moment when you don't know which direction things will go."

I knew that he was right; both planting and fire were necessary, and it was hard to know which, in the end, would lead to more destruction or creation. I had lived that moment many times before, and now I faced it yet again. Sometimes loss makes room for new ideas, new lessons, and new life—despite the painful sadness. This seemed to be the case in my marriage, and with grace, it would also be true for the reforestation site and for the planet.

21

Shedding Deeper Layers

2014–2015

Our plans to launch new locations were put on temporary hold as I gave my full attention to constructing a medical center at ASRI. It was simply not flowing smoothly. The building had been designed with help from students and faculty at Georgia Tech starting in 2009 (thanks to connections and support from generous donors). They did three classes led by an Indonesian graduate student who brought students to Sukadana, then professional volunteer architects and engineers helped, and finally, it was refined by key Indonesian staff. However, we were really struggling with fundraising and kept hitting bureaucratic permit delays and corruption.

One village chief absolutely refused to give us a critical letter of support. The implication was that he wanted a bribe. For more than a year, we had been visiting him to try to cajole him, but no matter how important we told him this building would be for the health of the community, he refused. He was young and healthy, and apparently, he wasn't able to easily put himself in the shoes of someone who wasn't. Our goal was to build a facility that would be able to do operations, but he was not convinced those, or more generally improved care, were necessary. We were all at a total loss. If he didn't give us his letter of recommendation, we couldn't move forward with the permit.

Then one day, he came into our tiny clinic with right-lower-quadrant abdominal pain—classic appendicitis. We told him we were sorry, even though he might improve on antibiotics alone, he couldn't be treated here given that he might need to have surgery and there was nowhere within two hours where that could be done. We would have to refer him to the city. We gave him a first dose of intravenous antibiotics and some pain medicine and then sent him off in the ambulance. The incredibly bumpy road could not have been easy. When he arrived, the caregivers in the emergency room told him the surgeon wouldn't be back for one or two days and that the next day they would give him an X-ray to determine if he did have appendicitis (side note: an x-ray is useless in diagnosing appendicitis). No more antibiotics and no pain medications were given. The next day, he did get an x-ray, but the doctor said the quality was too poor to determine if he had appendicitis and the surgeon was still gone. He would just have to wait for the next day. But on the second day, his appendix ruptured, causing incredible pain. His family left "against medical advice" and brought him back over that horribly bumpy road to us. He arrived about 5:00 p.m., and after a rush of working to stabilize his vitals, we sat down with the family and told them there was a good chance he wasn't going to make it.

The next weeks were nail-biting, and sometimes, I awoke in the night wondering if we should try to drain the huge abscess we could see on ultrasound with a big needle. But each day on antibiotics, it shrank until finally we felt it was safe for him to go home.

On his last day in our clinic, I stood next to his bed and told him he would still need to be very careful until he was fully recovered, resting more than usual and working less. He looked up at me and said, "Kinari, I will write you that letter."

I smiled, cocked my head to the side, and slowly nodded. "Yes, you will."

His illness had brought him around to what he knew in his soul he needed to do. I could empathize.

But the truth was, even now, I was still not in full alignment. I had been traveling like crazy to help raise the money for the new building as well as spending time in Sukadana navigating permit problems, hiring contractors, and caring for the occasional recalcitrant letter writer. All the travel and stress were proving to be too exhausting, and in the beginning of the third year after the jellyfish sting, my health began to collapse again. In addition to the overwork, I was also finding that when I struggled spiritually and emotionally, my energy drained much more quickly. While everyone experiences this, in my case, the effect was much more intense, because the toxin had damaged my spinal cord in the critical part where the brain regulates the nervous system. There was no tamping down of the somatic expression of the stress I was feeling. First, I began having trouble breathing, and I coughed intensely with exertion. Sometimes the coughing was so strong that I would have to curl up on the floor until a spasm subsided. When I wasn't traveling, my home base had become the spare room of my dear friends Patricia Plude and Steve Kusmer in San Francisco. I kept getting worse until it got to the point where I could not climb a single flight of stairs without Steve pulling me up. My pulse raced at random, and my blood pressure dipped so low, I had trouble standing.

Part of me knew Ida Resi was right and that the real problem was that my heart and my mind were still fighting. But rather than face those issues, I decided to look at the problem from a purely medical perspective. Maybe it was time I stopped being my own doctor. I sought treatment at Stanford University. Right off the bat, they put me on about five medicines. Then after extensive tests and a steady deterioration, my doctor sat me down. "It's probably going to just keep getting worse, and it very likely will kill you—although we might be able to delay that a little. I want to start treatment as soon as possible." He told me he thought the jellyfish sting had triggered an autoimmune response, and my only chance to slow it would be to knock out my bone marrow with chemotherapy or do weekly plasmapheresis (your entire blood is filtered and just your red blood cells are returned mixed with saline). Eventually, he thought they could taper the plasmapheresis down to once a month to keep my immune system basically deadened, but that that would probably continue for as long as they could keep me alive. Needless to say, any part of a life in Borneo or replicating our work around the world would be gone with either option.

With each glimpse of the sun, I was forced to shed more attachments and fears. I had already let go of seeing myself as primarily flawed. After being wounded, I had stepped permanently away from my role leading the little band of rebels who were the ASRI team. I had ended my marriage to Cam, moved away from my home of ten years, returned to the United States, and was exploring new ways of being in the world. Now it felt as if even deeper layers of myself were being wrenched off—whether I was willing or not.

When Cam had asked me to be brave enough to ask what *I* wanted, with extreme effort, I had managed to do it. But it was not yet a way of being for me. I was coming to realize that listening to myself had been so hard because of deep childhood patterns of suppressing any of my own needs. Now these patterns had to change with my family and in the way I interacted with myself. I had to be able, in the smallest to the biggest things, to listen to my own body and my own desires. I could no longer take care of my family and not consider my own needs. If I didn't change, I would literally die.

Another layer of loss was letting go of my dream of having a child of my own one day. After leaving Cam, I was nearly overcome with the desire to have a baby. I had tried seven rounds of intrauterine insemination with donated sperm and had been unsuccessful. It was a blessing it didn't work, as I was truly too sick to care for a child and didn't have the financial resources either, but it was also a sad loss. And now, Stanford physicians were telling me my illness was likely fatal. I almost wanted to shout at God, "Seriously? Do I really have to go through this again? Haven't I let go enough? Didn't I *already* have to face death multiple times?" Apparently not yet well enough. And as more layers slipped away, I realized that another core attachment was also at risk: my understanding of God. Because if death was imminent, that would mean that I had mistaken the calling, the dreams, and the message that came through Ida Resi. Yet if I was honest with myself, in my view of God "forcing" me, there was also resentment. I had grudgingly accepted, but not without fear and annoyance. I just wanted God to call someone else. Which if I was going to die, appeared to be happening. But with that, I also felt sadness.

So I let go of the calling, too—and with it, how I thought of the Divine.

My Mennonite pastor, Sheri Hostetler, had written a poem called "Instructions" that kept playing through my mind during that time:

> *Give up the world; give up self; finally, give up God.*
> *Find god in rhododendrons and rocks,*
> *passers-by, your cat.*
> *Pare your beliefs, your absolutes.*
> *Make it simple; make it clean.*
> *No carry-on luggage allowed.*
> *Examine all you have*
> *with a loving and critical eye, then*
> *throw away some more.*
> *Repeat. Repeat.*
> *Keep this and only this:*
> *what your heart beats loudly for*
> *what feels heavy and full in your gut.*
> *There will only be one or two*
> *things you will keep,*
> *and they will fit lightly*
> *in your pocket.*

So I watched each thing go: people I loved, roles I felt called on to play, my identity, God, and then life itself.

I almost just gave in and underwent the Stanford treatments—which would likely have killed me of their own accord. But before I did, I was strongly encouraged by my friends, including Patricia and Steve, to get one more medical opinion. I went to the best: the Mayo Clinic.

It's odd that the most renowned clinical care in the United States, and possibly in the world, is in the middle of nowhere in rural Minnesota, but there it is. Our board member Clare Selgin flew out from the East Coast to help me travel (I never would have made it without assistance), and my friend Rhiya Trivedi met us there.

It is amazing what accurate tests, a correct diagnosis, and stopping inappropriate medications can do for a patient. The magic came from an entire team of health care providers working respectfully together. It was not so much the doctors as the whole beautifully integrated system that made the difference. I had learned about the Mayo Clinic's remarkable approach in medical school, and I had actually built ASRI's systems along the same philosophies, including having daily consultation meetings with the doctors. My dream for the ASRI medical center was that it, too, would become a center of both healing and teaching. The Mayo's genius was in recognizing that all the parts of a patient's care needed to work seamlessly together, rather than in segmented specialties. It shouldn't be unusual to think of humans as whole integrated beings, but in Western medicine, it sadly is. At ASRI, though, we were trying to take it even one step further. Not only did we provide holistic medical care, we also recognized that a person's well-being couldn't be separated from their environment. We were expanding our circle to caring for nature and livelihoods as well. Funnily enough, I was even a good case example of how damaged ecosystems can hurt people's health, since jellyfish have been steadily increasing with the warming oceans and the loss of predators like turtles.

The Mayo Clinic's assessment was that I probably would *not* die from the autonomic dysfunction, but I would almost certainly die from the treatments proposed. Off the inappropriate medicines, my health began to improve within days. The "cure" of a proper diagnosis and hope surely helped as well. The process reminded me of how I felt one night in the rainforest when my flashlight ran out of batteries halfway home from following an orangutan. The forest had been pitch-black, and there was terror in not knowing how I could possibly get back to my little hut and that I was unlikely to survive the night. But as my eyes slowly adjusted, I realized that the faintly glowing phosphorescent fungus I had often noticed from my platform covered the entire forest floor—except for the path. The lack of leaves made a ribbon of pure black through the distant galaxy starlight of the forest floor. I followed the darkness, and after many excruciating hours of tentatively making my way along that fearful journey, it led me to home and safety.

I had bought only a one-way ticket to Minnesota, not knowing how long I would be there. Then feeling so much better, I decided on the spur of the moment to accept Rhiya's invitation to come join her in New York City where she was now in law school to pursue her passion for social justice. I sat on the plane and put my headphones in. "Gloria" by Michael Franti came on, with the refrain, "I'm glad to be alive," and it was everything I could do not to sing along full throttle.

Walking the streets of New York, I realized profoundly what a gift life was and that I could create it however I wanted. It was my *choice,* and I could bring back into my life whatever I wanted. No obligations. I could say yes, but only if I wanted to. And I did want to. I accepted the calling to serve as a "guardian of the tree of life" and to spend my life working toward planetary health. My mind was no longer resisting what my heart knew was the right path.

22

Finding Alignment

With my stabler health and internal alignment, the Health In Harmony team and I managed to finish raising the funds for the medical center. Two amazingly generous gifts made it possible: one from the A'lani Kailani Blue Lotus White Star Foundation, and the other from a lovely woman in her eighties named Betty Gardner who was a friend of my father's. That first gift happened right after I left New York to visit New Haven and the second when I went to visit my father in D.C. immediately after that—little miracles telling me that with alignment, the energy would flow. All told, with the continued support of many individuals and organizations, the $2 million for the building and our program operations came together.

After returning from the East Coast to San Francisco, there were two more steps that needed to happen. The first was switching to eating fully vegan. This process began after visiting Ida Resi and simply no longer being able to ingest so much suffering, but in studying it, I had found it also improves longevity, saves Brazilian rainforests uses much less water, and can even improve sports performance (see the film *The Game Changers*).

The second change was an outward one. When I had been in Bali and did the days of fasting and prayer that had led to a declaration of interbeing, there had been clarity that the three colors that would symbolize a commitment to the earth were green, blue, and white. I had known for years that I should shift my clothing to those colors, but I kept finding ways to justify not doing it. Now I knew I had to. So I invited friends over and gave away every item of clothing not in those colors (except for a little bit of red and hot pink for days of fun). And as usual, my fears of scarcity were unfounded, and I had more than enough with a few secondhand items added. Now, outwardly, there was alignment to my inner peace.

In August 2016, about a year after we started construction, I made another trip to Sukadana to oversee the final stages. I had moved to being an ASRI board member as well as a Health In Harmony staff member. I was the only non-Indonesian on the board, and the staff were now 100 percent Indonesian. I found such pleasure in walking through the nearly completed building and chatting with the neighbors who had been hired as part of the crew. Some days, I joined the ASRI medical team in imaginary walk-throughs to plan the patient flow, administration, and equipment.

When we had designed the building, we had shown community members various potential designs. My favorite comment was from a local woman who said that what they liked about the ASRI clinic was that it appeared to be a "normal" house from the front, though it was fancier and bigger on the inside. She said it made them feel comfortable to go there. And, indeed, we had made the new building look inviting and small from the front, but it stretched far back and was fully functional. We also incorporated many design features of local architecture, such as wooden shutters and wide-open porches like the butterfly porch.

During this time, I dreamed a comforting dream. I was stringing together a necklace of the most gorgeous malachite beads—twelve of them, in brilliant blue and green. Each of the beads seemed to represent other dimensions, and as I strung each bead, I was also experiencing those other layers of reality. Each bead was a person I had loved intensely in my life, or a place where I had experienced a powerful moment of change, or someone (still unknown) whom I knew would become critical in the work. As I arranged the necklace, another layer of reality emerged: the necklace became the medical center courtyard itself. This felt exactly right—that this welcoming space was created out of beads of love and experience and that it held the seeds of the future.

I was aware in the dream that I didn't know the proper order of the beads, but there was no stress. I was confident that each bead would eventually fit comfortably into its place of understanding. And when it did, energy would begin to flow around the circle like the healing and love that would flow around the courtyard itself. This dream felt in some ways like the answer to the one I had had when I was still struggling to find the right location for this work. In that dream, I had to reach through a kind of latticework mesh to crack a complex code that I did not yet know. Now, I reflected, maybe the code I wanted was simply this sequence—of people to love, changes to experience, and possibilities to create. And maybe the real prize was the magnificent medical center and all that it represented for the planet.

The whole building was also designed so that nature was ever present. The main clinic area was built around a landscaped courtyard, and each room looked out on trees. We had a second-story open meeting area with views onto the national park behind. From the rooms in the morning, the gibbons could be heard calling from the hills. As a planetary health healing center, we wanted people to fully understand the interconnectedness of nature and health, and studies show that if you can see trees, or even just green space, from a hospital room, you will recover more quickly and need less pain medication.[1]

---

1 R. S. Ulrich et al., "View Through a Window May Influence Recovery from Surgery," *Science* 224, no. 4647 [1984]: 420–21.

We wanted the building to be built as ecologically friendly as possible. The contracting company, Pilar, was impressively accommodating of all our requirements. We insisted on knowing where every single load of sand, every piece of wood and bamboo, and every rock came from. Our conservation team members were out in the villages checking on sources, approving or rejecting each one. We also required Pilar to hire as many local laborers as possible—especially loggers. These loggers and other neighbors were not only earning money but also learning new, marketable construction skills. One of our best local workers was a man in his forties with an unrepaired cleft lip and palate who kept telling me in his slurred speech how hopeful he was that someday he would get surgery in the building he helped build.

Behind the building, we planned a large seedling nursery to care for the seedlings patients brought, and an organic garden that would feed the staff, be a place for family members of patients to work, and allow us to test new techniques we might want to teach farmers.

One day, the former village chief of the Sukadana area joined me to walk along the construction site. He turned to me with moist eyes. "Could we have imagined, when we designed this program, that we would ever have something as amazing as this building? Health is the foundation of *everything*. Without it, you can't achieve anything. But with this building, nothing is going to stop us!"

AS BUILDING NEARED COMPLETION, MONICA and I decided to do a round of radical listening evaluation meetings. It had taken us some time, but after ten years, we had essentially met the needs that the community had requested. Now we wanted to know if solutions for protecting the forest had changed.

During our first meeting, Monica and I just stared at each other across the circle, dumbfounded. They wanted more reforestation. We were even more surprised when this result played out in all the meetings we held. One old woman explained, "You have to understand, this isn't just for us, it is for our children and grandchildren." These communities were far from wealthy, but recovering the health of the forest was now their primary desire. They wanted us to plant in many more communities, not just a few. Secondarily, they also requested more help with eyeglasses. I had recently heard that the eyeglasses company Essilor was interested in a corporate social responsibility partnership, so this could dovetail nicely.

I had been quite depressed about reforestation after our devastating fire, but it turned out there was more resilience than I'd expected. Thirty percent of the plantings had resprouted, even though all of them had looked completely dead at the time. These new sprouts were able to grow very rapidly, as their roots were already established and they were getting lots of sun. As requested by the community, we had also gone back in and done enrichment planting. Some areas already had a fully closed canopy. There had been a dip in the bird species after the fire, but the number was now even higher than before the fire at seventy species.

One of the best learnings came from the ten-by-ten-meter control plots that Cam had put in and left unplanted for comparison. Even after six years, in the case of the oldest ones, they were still just scrubby grass and ferns with no trees at all. They looked like empty divots amid the replanted forest that towered over them on all sides. The lesson was clear: just keeping out fire was not enough. These grasslands would require active reforestation to bring back the trees. And that is what the villages around the park now wanted: "Please help us bring back the forest." Apparently, they understood what Cam had shown in his experiment.

IN 2017, WE COMPLETED A ten-year survey to understand the ongoing impact of our work more analytically. One exciting finding was that infant mortality dropped 67 percent from our baseline (with a concomitant 55 percent reduction in the birth rate). And equally exciting, we found logging households had now declined 90 percent. This calculated out to now an estimated 150 loggers, down from the originally approximately 1,350 logging households. The Forest Guardians counted all the loggers they knew about and got to 160, providing independent confirmation that loggers really had dramatically dropped.

However, we all—including the communities—wanted to get to zero. To achieve that, Dr. Monica and the conservation team, in consultation with the community, came up with a creative solution. One of the unusual things about these last holdouts was that most of them didn't own enough land for farming, and they almost all owned their own expensive chain saws, representing either a large investment or inheritance. It was decided that we would buy the chain saws and also provide additional start-up capital for these families to start their own businesses. It was basically angel investing without interest. We required both the husband *and* the wife to start a business—or they could go into business together. We provided business training for them both. The businesses were jointly owned with ASRI until they paid off what we had invested.

It was impressive how successful many of these microenterprises were becoming. One family was raising chickens; the husband cared for the fowl, and the wife sold the meat from the back of her bicycle, riding around town with a cooler strapped to the rack. Another couple realized a long-held dream: they started a barber-and-boutique shop, where the wife rented wedding outfits and the husband cut customers' hair—instead of trees. And another man started a café and karaoke bar called Stihl Hot, after the brand of his chain saw. Interestingly, he didn't actually know that the name worked well in English. He chose the term *hot* because that is what people would say if someone is particularly great at karaoke—not because the coffee was "still" a nice piping-hot temperature.

Our plan was to use the chain saws to build a number of sculptures that would look like coconut palms. The long blades of dismantled chain saws would be bent to look like leaves, and the bodies would make up the trunk. Those chain saw statues would stand as monuments to

ASRI's success in ending the destruction that used to occur. One would go out front of the medical center, one at the national park office, and possibly others would be installed at the offices of the government and academic partners who had helped us achieve success in Indonesia. But the most important one would go in the middle of the courtyard of the medical center. ASRI's patients and people from all over the world learning to replicate this model in many other places would be able to hear orangutans' long calls and the musical songs of birds while look-ing at a visual representation of what used to be, but no longer was.

We still needed to know if this drop in loggers was translating to the loss of less primary forest, even if forest was starting to grow back. Stanford University helped us by doing an independent assessment using satellite imagery and comparing Gunung Palung with other national parks in Indonesia. What they found was impressive. While most national parks continued to lose ancient forest during the time ASRI worked, Gunung Palung barely lost any. The difference was $65.3 million worth of carbon.[2] The world had given to these communities about $5.2 million worth of thank-yous (including the medical center), but the gifts that were given back to the world were worth many times that.

In addition, that carbon figure above didn't count the logged forest that was growing back, which satellite imagery showed to be about fifty-two thousand acres.[3] Nor did it count the few hundred acres we had planted in areas that never would have grown back on their own (the impact compared with the natural regeneration was minimal but still accounted for the equivalent of taking forty-two thousand cars off the road). Over the years, it had been a joy to see the hills around the communities that had been so badly degraded slowly turn green and have full streams flowing down from them.

---

2 Jones et al., "Improving rural health care reduces illegal logging and conserves carbon in a tropical forest," *PNAS* 117. 45 [2020], 28515–28524.
3 N. I. Fawzi et al., "Forest Change Monitoring and Environmental Impact in Gunung Palung National Park, West Kalimantan, Indonesia," *Jurnal Ilmu Lingkungan* 17, no. 2 [2019]: 197–204.

It appeared that our work after ten years really had contributed to community and global thriving, and I knew now the time had come to expand its impact. Finally, I was both physically and emotionally ready. My heart and my mind were one.

23

Lemurs, Jaguars, and Hornbills

2017 AND BEYOND

O ver the twenty years I had been paying focused attention to climate science, the predictions had become steadily more dire. At first, it was something that might affect grandchildren, then children, then suddenly it was happening *now*. Our sense of urgency really increased when, in 2018, an Intergovernmental Panel on Climate Change report (IPCC) warned that if humanity didn't halve the annual emissions produced by 2030, we could go over irreversible tipping points. In 2019, eleven thousand scientists affirmed that assessment and predicted "untold suffering" if positive feedback loops led to a hothouse earth. Yet with all the press, I still found it remarkable how few people realized that we are literally looking at an unlivable future for almost any life on earth. Most of the wealthy people I knew somehow imagined their money would protect them—but that simply isn't true. If food can't be grown, privilege is worthless. *The Guardian* quoted climate researcher Dr. Phil Williamson as saying, "In the context of the summer of 2018, this is definitely not a case of crying wolf, raising a false alarm: the wolves are now in sight."

Given the lack of successful conservation models, we felt we had an obligation to see if our model would work outside Indonesia as well, and we realized we were going to have to move incredibly fast. However, I knew I did not have the skills to build a multinational nonprofit. Michelle Bussard, in her wisdom, also knew this was not her forte, and she decided to step away to make room for someone else. The board under the brilliant leadership of Jo Whitehouse hired Jonathan Jennings to lead Health In Harmony. He came to us from his role as deputy executive director of Doctors Without Borders in Canada. He had managed medical humanitarian relief work in South Sudan, Ethiopia, India, the Democratic Republic of the Congo, Uganda, Liberia, and Kosovo. Jonathan was clear that he no longer wanted to take care of crises caused or exacerbated by the climate emergency without actually addressing that root cause. One of Jonathan's requirements for taking the job, though, was that I would stay on and assist him in vision, fundraising, and spreading the word.

With the medical center's ribbon cut, brilliant new staff on the Health In Harmony team, strong leadership at ASRI, and excellent boards in both the United States and Indonesia, we prepared to begin our second site. One aspect of this was determining what we were actually replicating. What were the key elements of our model? We thought they were: 1) rainforest community design of solutions through radical listening; 2) precise implementation of those with a long-term commitment; and 3) addressing systems problems in integrated ways that make the linkages visible both locally and globally (like the noncash payment options, incentives for non-logging communities, and education on both health and environment). However, there were some things that we had done unintentionally that might have been very important. For example, we had mostly had female leaders and empowered women throughout our whole program. Was that essential or not? One of our community health workers, Hamisah, whom we had promoted to our TB coordinator (she reduced the dropout rate to an unheard of 0–2 percent), had even become the first female village chief after saving so many lives in her village. Riane Eisler's work on dominance versus collaborative societal systems would suggest that female leadership could be critical. In dominance systems, the goal is to "win," which means *power over.* The socialization of this pattern begins in childhood with gender relations but is transferred to racial interactions, religious ones, and ultimately the way we treat the natural world. In contrast, collaborative systems, which, she argues, occurred for tens of thousands of years of human history, emphasize *power with.* These societies have more egalitarian gender relations. Interestingly, Lynnette Zelezny and colleagues reviewed studies from around the world in 2000 and found that women are likelier to act pro-environmentally than men, so maybe female leadership really did matter. We would have to see if we could figure it out.

We also suspected it was important that we were framing the agreement with the communities as mutual gift-giving. Behavioral science shows that when we do something from our own love, compassion, and desire for something to happen (intrinsic desire), not because we are being paid money to do it or because we are trying to avoid punishment (extrinsic), then humans are much more motivated to accomplish it. Once something is monetized, it can be extremely difficult, if not impossible, to go back to heart motivation (this explains part of my aversion to the carbon markets). We can lose something important if we simply put a dollar number on these forests instead of recognizing their beauty and inherent value and giving gifts to support those who can protect them.

Our model was also one that would work mostly in areas where the primary threat to the forest was the communities themselves—not big agribusiness or mining. A helpful paper in *Science* in 2017 by Alessandro Baccini and colleagues showed that this is not a rare situation. They looked at forest density from satellites instead of just forest cover (in this way, you can account for individual trees being lost), and they found that 69 percent of the loss of carbon in tropical forests is actually due to the degrading of forest through individuals logging or fire. This is not to say that total clear-cutting for something like palm oil is not important—it certainly is.

It is much harder to get back to intact forest from clear-cut land than from logged forest. But if all the conservation programs focused just on agribusiness, we would be missing 69 percent of the problem.

After our initial scouting trips, the ASRI and Health In Harmony teams had completed other assessment visits to two other sites in Borneo and also started to look outside Indonesia. We decided our second and third sites would be in Bukit Baka Bukit Raya National Park in West Kalimantan and Madagascar but were open to adding a fourth site—probably in Brazil.

In 2018, we started working in Bukit Baka Bukit Raya, a park nearly twice the size of Gunung Palung. The Indigenous Dayak communities there had asked for midwives and help with their agricultural practices in our initial radical listening meetings. These communities had been headhunters within my lifetime and still lived very traditionally—except most families now had chain saws. One woman declared that if anyone denied they had logged in the national park to pay for health care, they were lying to me, because, like at Gunung Palung, that was the *only* way to get enough money for medical services. And, indeed, in our baseline census, 84 percent of households admitted illegally logging to pay for health care. Before we placed two midwives in the nine villages where we began work, most adult women were traveling four to five hours round trip each month to get birth control, and the cost of this was five planks per month. But, again like at Gunung Palung, our baseline survey showed that nearly everyone (94 percent) wanted the forest to be there for their children and believed it needed protecting (98 percent). They just couldn't do it on their own.

One amazing thing about providing health care as a gift to help protect the forest is that it engenders incredible trust—possibly more than anything else could do. To honor their cultures and traditions, we made sure to hire two Dayak midwives from the same region. These women almost immediately started saving lives—as well as making life *much* easier by doing simple things like providing birth control. However, they also ended up treating malaria, typhoid, and even setting broken bones, so after about eight months, we hired a doctor and a nurse as well.

The energetic shift in these communities was profound. They had moved from constant fear to a state of eagerness and calm. One of the midwives had even organized the women to sweep the village every three days, and it nearly sparkled. (Each sweeping day was followed by getting all the women to do Zumba to solar-powered music in the streets for exercise.) The Forest Guardians and community members we had sent to the island of Java for training in organic farming techniques were also already helping people to shift away from logging.

With Bukit Baka Bukit Raya off to a great start, we began work around Monombo Reserve, on the southeast coast of Madagascar. Madagascar had lost 80 percent of its forest, mostly from individuals cutting down trees or burning them to meet their basic needs. In 2019, I went to hire our first Malagasy staff and confirm the radical listening solutions the team had heard on our first assessment trip.

Having lived in a little village in Borneo for seven years where the average income was less than two dollars a day, I wasn't particularly fazed when people told me Madagascar was poor—I figured I knew poverty. But the fourth-largest island in the world is nothing like the third largest—the poverty is extreme and nearly ubiquitous: adults who don't even own flip-flops, entire communities where not one person has been able to afford a bicycle, and almost no electricity for most of the country. Madagascar officially ranks among the poorest countries on earth with 80 percent of its twenty-five million people living in extreme poverty. Horribly, that truth has led to the near total loss of the forests, which are some of the most biodiverse ecosystems in the world, with 90 percent of the species occurring only in Madagascar, including many enchanting species of lemurs and even bioluminescent chameleons. As everywhere in the world, when basic needs for health, financial stability, safety, and dignity are not met, it's essentially impossible to follow through on one's desire to protect the natural environment. Social justice and environmental justice throughout the world, including in the United States, are inseparable.

People told us in radical listening meetings that now they knew that the saying of their ancestors that "there would be forest until hens have teeth" was not true. They could see with their own eyes that if they continued making charcoal and clearing forest with fire to plant crops of rice, they would soon have no forest left. But before we came, no one had ever asked them what the solutions were.

What they said was powerful: hunger and the lack of health care were the key problems. When I asked which of these two problems was more important, one woman said, "It is impossible to rank them. They are both number one. If you do not have food, you cannot be healthy, but if you have food and no health care, you will still die." Ideally, they wanted mobile clinics to centralized villages. Walking four hours to a road and then waiting for the Taxibus, which was usually already full, was simply not possible when one was very sick.

To address hunger, they needed help shifting fully to rice agriculture in the marshes outside the protected areas. With improved irrigation systems, three-month variety rice, and modern rice-farming techniques, their yields could improve. However, they told us they would need to be paid in food to improve the irrigation systems, because without it, they would not have the energy to do the labor. They had essentially gone over a tipping point—they no longer had the energy to do the work they knew needed to be done.

Their third request was for help rebuilding their schools. In the distant past, the government had built these schools, but they were now crumbling, with holes in the tin roofs, cracked cement floors, and walls made of traveler palm fibers with large holes. One man laughed that someday they wanted their children to be paid just to talk into someone's ear like the young man next to me. Tojo, a young man from the local university who also worked with tourists, blushed as he translated and then even more so as he had to translate back my praise of him.

After these gatherings, the community members told us they were the best meetings of their lives. A friend of ours with extensive experience in Madagascar was traveling with us, and I asked her if these effusive responses were typical. She said she had never seen anything like them. Soon after that visit, we signed agreements with the village "kings" and began to address the solutions they had outlined.

IN 2019, I ALSO WENT to Brazil to evaluate if our model might work in the biggest rainforest on earth. I was horrified by what we saw: thousands of acres of forest (enough land for hundreds of Indonesian families to live on) could be cleared by *one* family for cattle. Watching the cows graze around stumps of enormous trees reaffirmed my own decision to go vegan. I was more appalled when I discovered that these families were not making large profits by cattle ranching. In fact, they were often losing money.

It made even less sense when we visited fruit farmers making up to $75,000 annually on just twenty-five acres. Clearly, something else besides economic maximization was going on if growing native fruit trees like açaí, cupuaçu, and cacao on average yielded ten times more profits than cattle ranching. At a dinner with a scientist who studied cattle ranchers, the situation became clearer: her research found that the primary reason for this decision was what they called "safety"[1]: cattle were basically a walking savings account that they could use if they had an emergency. Fruit, in contrast, only came during its season. People who had recently decided to clear land told us that the key element of safety they needed was around health care. While Brazil had free universal health care, transportation across the vast landscapes, living in the city during treatment, and lost income could easily add up to very high costs. Long waits for the under-resourced public system also meant

1 R. D. Garrett et al., "Explaining the Persistence of Low Income and Environmentally Degrading Land Uses in the Brazilian Amazon," *Ecology and Society* 22, no. 3 [2017]: 27.

sometimes people just paid for private services. When I asked the fruit growers how they got around this, they told me that they had pooled money for medical emergencies in collectives.

We also traveled deep into the Amazon and spoke with traditional and Indigenous peoples who guard the intact forest and live off what the forest produces. What we heard over and over was that lack of access to health care and the incredibly high costs of getting to the city for it (let alone the extreme difficulty) were the biggest threats to them staying in the forest and continuing to protect it. It is easy to understand why the Brazilian health ministry would make a utilitarian decision to spend its resources in the cities and save more lives. But providing health care to these remote regions of the Amazon would be cheap in terms of protecting the whole planet. Interestingly, we also heard in a radical listening meeting that they would ideally like to mix Indigenous and Western medicine, taking the best from both. This solution would also reduce costs and side effects from medications.

Our hope was that through partnering with a local Brazilian non-profit like Instituto Socioambiental (ISA) we would be able to address dual threats to the forest: cattle ranchers invading from the edges, and Indigenous and traditional peoples leaving because of lack of access to health care. We ideally wanted to work in the Xingu basin, which is the size of the United Kingdom but has a population of only twenty thousand, compared to the UK's sixty-six million.

IN ADDITION TO TESTING OUR model in new sites, we also knew that part of scaling would be to teach others how to implement radical listening and be a part of shifting the cultural paradigm about what is possible. As well as teaching radical listening virtually, we had designed our new building to also be a training center. One of the village chiefs described his community as "pathfinders for how to live in balance with the natural world," and he said they now wanted to teach the world. We wanted to support them to have the opportunity to do that. Our vision was to provide individuals, other nonprofits, and even government employees a hands-on course in how human and environmental well-being are intertwined so that they could learn from these communities.

We were delighted when, in 2015, we finally had a term for what it is we had been doing: planetary health. The Rockefeller Institute and *The Lancet* coined this term, although the concept has been in practice in various Indigenous communities for thousands of years.[2] The term nicely encapsulated how interconnected all life is on earth in a way that most lay people could understand. We started calling ourselves "Planetary Health in action." Each year, we began to have hundreds of students come to learn from the ASRI staff and the communities around Gunung Palung.

Still, we knew that replicating our model and teaching others to do so, too, likely wouldn't be enough. Like in my prophetic dream, I knew that we needed to have those back channels of communication based on Indigenous and local wisdom. What I was envisioning was a technological means of connecting the radical listening needs of communities with global citizens who could support the organizations that would meet those needs. I envisioned uploading to a societal platform the radical listening solutions of communities all over the tropics in how to protect forest and improve human well-being. I also wanted local communities and global citizens to be able to see nearly in real time the impact of their joint desire to protect forest through satellite imagery. In this way, global citizens, companies, governments, and foundations who had access to resources could meet those needs. Given the urgency and the scale of the problem, this was the only way that I could imagine change happening at the speed it would need to. I was also painfully aware that if we just wait for our governments to solve these prob-

---

2 Redvers, Nicole. 2018. "The Value of Global Indigenous Knowledge in Planetary Health," *Challenges* 9, no. 2: 30. https://doi.org/10.3390/challe9020030.

lems, there likely really is no hope. This is not to say top-down policy changes wouldn't be extremely useful, but as citizens of the world, we will have to find a way to heal our planet—even if governments are actively working against us.

In my prophetic dream, I had had to see the reality of the poisoning of our earth and accept that I must fully agree to fight in the great battle of our time. Possibly, the greatest war of all time: the war for a survivable planet. In the dream, once that acceptance had happened, the children ran to the valley to collect all those who could help coordinate the resistance. And indeed, at the time of writing, the children of the world are rising up and demanding that those of us who are older start acting like adults and protect the children of this earth.

For those of us who have heard the message of these wise youth, now comes the time of working together. Each of us has a critical role to play. Every single person on earth has something to offer to the solution. Though I would prefer that my dream did not use metaphors of war, maybe these are the ones that we can most deeply understand as a human species. But in this war, the nonviolent weapons are water balloons of truth and love. And we cannot lose, because we do truly hold the moral high ground—we are fighting for all our futures.

24

Becoming Guardians

OUR FUTURE

Imagine for a moment with me what a positive future could look like. Imagine a world where everyone has access to health care and their basic needs are met. A just world where decisions are made locally that benefit people now and far into the future. Imagine forests and other natural ecosystems recovering and even expanding while people's economic well-being improves in a regenerative manner. Can you see rainforests regrowing in Indonesia, the Amazon basin, the Congo, the Philippines, India, and Madagascar? These forests would be absorbing and storing carbon while releasing oxygen and providing habitat for incredible biodiversity.

What if enough of each fishery was set aside for the fish to thrive and the catches to actually increase? I can see quiet cities with all electric cars, efficient public transportation, streets just for bicycles, excellent clean air, and abundant trees. And what if there were wildlife corridors across landscapes and even through cities? Imagine honoring and restoring critical habitat for bird and butterfly migrations across the world. And even dare to think about banning pesticides, allowing the insect population to rebound even more than the 80 percent they have dropped since the 1970s. I can envision restorative economies where everything that was made was created in a way it could be re-purposed, fixed, or completely recycled when we had finished using it. Three-fourths of the clothes produced should not end up in landfills or be burned. Imagine local production by creative artists rather than systems of exploitation. The crazy thing is that all these ideas don't create *less* for humans—they create more! More health, more resources, and more happiness. And there are people and organizations showing that each one of these ideas is possible. If we can envision a positive future, as Elena Bennett says in her brilliant TEDx Talk, we might just be able to create it.

But for us to get to a place of thriving, we are going to have to know enough-ness in our bones. We are all trapped in a fear of scarcity. Paradoxically, it is that fear of scarcity that actually creates the lack itself. We have to see how much we already have and be willing to share the overflow. My own scarcity complex was obvious in Sukadana. When someone was in need, the generosity of neighbors was incredible. From very little, they always gave, and shamed, I learned how to follow suit. Some of us have more than enough, and others are below the floor of basic needs being met. The truth is that there are also reparations to be paid. Many do not have enough because of hundreds of years of resources, lives, and freedom being stolen. But there are more than enough resources to give this restitution. It will look different in different places, and the process should be based in radical listening. I believe part of the healing from colonialism and racism is to trust that those harmed know what is needed and that the power and control should be in their hands. Reparations, whether for descendants of enslaved people in the United States, Indigenous people in the highlands of Peru, or low-caste people in India, will actually lead to more well-being for us all—not less.

I have learned in my journey that the work of global transformation has many layers, and our ability to work together requires the personal work of healing. In fact, without healing ourselves, we are likely incapable of healing the planet. It all goes together. The more that we can find our own strength and come into all that we can be, the better. This means bringing to light the wounds that often come down to us through our ancestors. It requires extreme bravery to let go of old ways if those patterns are not supporting us. Imagine the benefit to the planet when every one of us recognizes our own amazing capabilities and begins striding forward in power.

The stronger each of us are personally, the more able we are to work together to find joint solutions to systemic problems. This means being willing to listen, even when what we hear may not be what we expect, and it will almost certainly involve facing our own privileges and biases. The humility to know that the best experts are those who are experiencing a problem is also critical. Indigenous knowledge of connection to place and decision-making about what will affect seven generations in the future must be prioritized.

This is the time when we must all learn to work together and coordinate so that our runaway horse doesn't go over the edge of the cliff. Personal growth is not enough. Personal healing and healing our communities is an iterative process, and the work in each drives the learning in the other. Progress in both spheres creates confidence that change is possible. Imagine how empowered we would all feel if, like the communities around Gunung Palung, we were *living* positive change happening.

How might things be transformed if each of us were willing to make a deep commitment to this change—a personal commitment, in dedication to each other and to the earth? What if each of our hearts and minds stopped fighting and the message flowed from our mouths?

This will be a huge struggle as we see old belief systems of scarcity intensely resisting change, but I believe that humans can achieve a profound evolution in our relationship to each other and to our earth. And I believe that those who resist it will, in the end, be grateful they failed. Each of us will have our own special role, and I know that we are *all* called. May those who need to hear it receive the message that healing is needed for our planet. May we all understand that, through profound listening to ourselves and others, we *do* have the capacity to achieve this.

Will you join me in this great work of our time and sign the declaration below?

DECLARATION OF INTERBEING

I hereby declare my intention to work toward a healthier and more sustainable relationship to the planet on a personal, community, and global level. I will become a force for healing and change in the face of the greed, injustice, and destruction that threaten the viability of the earth for sustaining human and nonhuman life. I recognize that we have little time left to switch from a path of destruction to one of thriving. I know we must all work together collaboratively to bring about change quickly.

I pledge my effort, resources, skills, knowledge, and desire. The tools of this work are nonviolent reorganization, love, radical listening, wisdom, equality, justice, compassion, innovation, determination, persistence, and truth. I pledge to work together across nationalities, religions, cultures, social classes, genders, and identities, knowing that only in diversity will we find enough strength and wisdom.

I recognize that even the smallest actions—both internally and externally—can have profound impacts. This is not work of self-denial but rather of enhancing and improving our lives individually and collectively: knowing there will be enough.

The colors of this work are blue, green, and white, as a symbol of being united in these goals. The blue of sky and water and the green of the planet's rich vegetation honor our parent the earth, in whose name we work. The white of clouds and snow is a sign of our willingness and desire to be visible. And, as in many cultures, the white also represents the death of an old way of being.

I sign here to indicate my commitment. My signature is a sign of the depth of my conviction, my desire for change, my willingness to work together, and my readiness to promote human well-being and the sustainable health of the natural world.

SIGNED,

Resources

If you would like to add your signature online, please see the website below. The website also provides further resources about personal and global transformation and suggests concrete steps you can take.

guardiansofthetrees.org

If you are interested in finding out more about Health In Harmony and Alam Sehat Lestari (ASRI) and to contribute to this work, please see:

www.healthinharmony.org
www.alamsehatlestari.org

Afterword

I would just like to note here that this memoir is a journey through time and reflects my own evolution of thought as well. I have been working for many years on understanding my own unconscious biases and the ways that white privilege and colonial privilege have blinded me to seeing the perspective of others—however, what I have mostly learned is how much I have to learn and that this will be a lifelong journey. I have always tried to approach this work from an anticolonial perspective, but I am also aware that there are aspects of our work that inherently have colonial overlays. We are working actively to try to combat those, and some of that journey is not reflected in these pages.

Given that the book is a memoir and is intended to show some of the evolution of my own thinking over the years, I left in things I did and said that now make me cringe (although I imagine my future self will cringe at my current self's understanding). I am also sure there have been ways that I have neglected to give full credit to what I have learned from the communities that I have had the privilege to live with. I realize as well that some of the imagery in my dreams may be seen as cultural appropriation and that I may have borrowed metaphors and ideas in selective or inappropriate ways. I hold myself accountable for the gaps and errors in this story, and for potentially referring to Indigeneity, Indigenous knowledge, and/or Indigenous peoples without proper context or cultural perspective.

I am also aware of my own cultural bias, which tends to value Western knowledge systems above other sources of knowledge. Part of my journey has been a process of learning that Indigenous and traditional expertise and science are equally important and bring great wisdom and knowledge.

To gain a deeper understanding of some of the key ideas in this book from a more academic perspective, please see the resources collected at guardiansofthetrees.org.

Acknowledgments

None of us are independent. We each exist in a web of support, love, and community. I feel that I have been particularly blessed in this, and below, I have mentioned just a few of the critical people who have made all my adventures possible and helped nudge me when I strayed. But I still get to take credit for all the mistakes. If I'd listened to these people more, I'm sure they wouldn't have happened.

For the writing of this book, I want to thank first and foremost Connie Gersick and Clare Selgin Wolfowitz. They tirelessly read and edited many versions of this book. But most important, they believed in me and the writing of it and kept encouraging me when I flagged. Tamarra Kaida, my first writing buddy in Bali, also was a huge help in the very beginning stages when I wasn't sure if I even had something worth sharing. And then I am grateful to my agent, Susan Golomb, who took another crack at editing and could see where the gaps still were and where it could be more concise. Finally, I had the total privilege to work with Sarah Murphy and Caroline Bleeke at Flatiron, and all I can say is that their work is pure genius. They could hold the whole picture in their minds at any time and see where through-lines needed to be emphasized, totally changed, or tweaked. Caroline slimmed the book and rearranged quite a bit, yet somehow managed to keep everything important and even make it much better. Thank you to all of you!

My greatest thanks have to go to my parents. They taught my sister and me both actively and by example that there are many roads less traveled—and the conventional ones aren't necessarily right. My mother also instilled me with her sense of beauty and taught me by example that I could do anything I put my mind to. I'm also very grateful to her for being willing to go through this hard process of healing together and in the end bridging the space between. Both my parents were serious about fostering our intellectual development, and my father was very conscious that raising girls in a male-dominated world meant that he had to actively teach us independence. For those of you who know my father, you will also hear in this book many of his thoughts that have helped me frame my understanding of myself, of my work, and of the world. Thank you, Daddy, for being the thousand-pound brain! My dad also connected me from the very beginning with his friends who provided the financial support to get the ball rolling—and then keep rolling. I always think of the Paul Simon song "Diamonds on the Soles of Her Shoes" when I think of his friends in D.C. I'm sure we never would have made it without his constant encouragement, support, and guidance. I also want to thank my stepfather, Jonathan Kingson, for being a steadfast support to the whole family and loving partner to my mother.

When my mom and Jonathan came to visit in 2008, Jonathan's comment upon arrival at our home in Borneo was perceptive and true, although I hadn't thought about it in that way until he said, "Wow, you had to travel to the other side of the world to find a place that was like the town you grew up in." Jonathan had moved to Dixon around the same time my parents did, and he and my mother married shortly after I went to college. It is likely he was referring to the close-knit community, the way people just dropped in without warning, but also the extreme funkiness and the way the natural world moved through everything (snakes, civets, and green-and-pink geckoes big enough to eat bats and mice). One of my dad's strongest principles was what he called "No Squeems." That meant squeamishness about anything was not tolerated. Just like me, when my sister visited, she only briefly blinked at giant spiders ("Never seen one that color before"), bugs ("No problem"), giant worms ("Oh, how interesting"), something dead ("What is it?"), or a rat ("Wow, look how big it is"). Out of all the volunteers who had been there, we were probably the coolest cats around. That is all likely thanks to the "No Squeems" rule.

On that visit, Jonathan gave Cam and me one of the best gifts we had ever received—an indoor sink and a sun shower. They actually checked a round metal sink into the baggage claim that showed off my mother's flair for design! But getting everything else was even more of an adventure while I was busy in clinic.

Despite not speaking Indonesian, every day, Jonathan rode a bicycle into town and got the next set of plumbing pieces he needed; it was a big game of charades, but in the end, he always got the right part or the closest alternative available. With his scraped-together elements, he set up an extremely basic gravity-fed plumbing system, where water was pumped to a tank on the hill and then ran down to the house.

I also want to thank my sister, who has been my dear friend and great companion through some really rough patches in both our lives. She inspires me and gives me faith in just how strong, smart, and loving a person can be. She has also been a fabulous mother to one of my favorite human beings on the planet: my niece, Clementine.

I've also had a string of adoptive parents who have cared for me over the years when I have sorely needed it: Susan Oberlander, Dagmar and Uli Steger, Margie and Wendell Geary, Jackie and Stuart Webb, Tamarra Kaida and Paul Knapp, Jo Whitehouse, and Pat Plude and Steve Kusmer.

And then there is Cam. How could I ever thank him enough for twenty years together (counting the time before we got married)? We each accomplished more together than we ever could have alone, and I pray that we will always be very close friends. In all the years of medical school and residency, Cam did essentially everything in our lives (cooking, shopping, bills, etc.) so that when I came home, we could have fun together. And then again, he cared for me so well when I was bedbound from the jellyfish. Cam also supported us both financially when for seven years I took no salary. And maybe most lovingly, he has been the best possible support as we transitioned to a life apart.

I also want to thank Ratna, Yani, Eka, and Komang, who have been our helpers at various stages in our life in Indonesia. Having help cleaning, shopping, doing the laundry by hand, cooking, and providing daily friendship support has truly meant that both Cam and I could accomplish so much more in life. We couldn't have done any of it without them.

And then there are all the amazing staff at ASRI and Health In Harmony. It was a joy to begin the journey with Hotlin and to go so far down the road with her. I am grateful for all she is, for all she taught me, and I wish her great joy and happiness in life. We have each transitioned away from the daily running of this program, but it will always be the child we grew together. She founded and is now running another nonprofit in Sumatra called Healthy Planet Indonesia that will continue the work there and hopefully help save the Tapanuli orangutan—a new species discovered not far from where Hotlin was born.

But we couldn't have accomplished anything without all the other incredible staff of ASRI. I want to particularly mention Wil and Clara, our first two nurses who have stayed with the program and will hopefully be there until they retire; Monica Nirmala, who took over running the program before leaving to do a master's in public health at Harvard; and Nur Febriani, who has taken over from her and is making the ASRI program even stronger. Mahardika Purba now runs the incredible conservation team. Ibu Lia has been with us for many years and is the key person who makes everything run at all—without her logistics support it would all crumble.

And then there were all the staff at Health In Harmony who get joint credit for making the work happen. Brita Johnson began the team, Michelle Bussard led the charge, then Trina Noonan held down the fort, before Jonathan Jennings took the helm. Trina has been the rock upon which it feels like Health In Harmony has stood from nearly the beginning. Ashley Emerson has also been an absolutely essential addition to the team in directing all the programs as we scale. I'm delighted to be on the next stages of this journey with the whole Health In Harmony and ASRI team. Please see below for a complete list of all the ASRI and Health In Harmony staff who are such amazing pathfinders for the world. It was a true partnership across the world, and it wouldn't have happened without the efforts on both sides.

I also want to thank the communities surrounding Gunung Palung who have been my most profound teachers and who truly are pathfinders for the planet.

We couldn't have done anything without all the people who have given of their life energy to fund our work. Unfortunately—and fortunately—Health In Harmony has a policy of donor privacy, although I would prefer to write every single one of their names here and thank them. If they had not been willing to partner with the rainforest protectors of the earth, we would all be the poorer for it. Thank you so much! Those mentioned in the book gave me explicit permission to name them, but there are many thousands more who put the fuel in the tank for the work to happen (or rather the solar power ☺)!

In addition to the folks mentioned in the text, there are a few key people who have also given me permission to give their names and some of whom were in earlier versions of the book. Tim Waters was so key in the beginning when he agreed to sponsor the salaries of two doctors for an entire year. Al Pierce and Lola Reinsch were the key donors who got all the architecture work done for the medical center by sponsoring the classes at Georgia Tech. Mimi Plumley has not only been a tireless contributor, she also did some very creative things like request sheets instead of flowers when she was sick in the hospital so she could donate them to us. This was critical because the quality of fabric one could get in Indonesia was just awful, and they wouldn't last many washes. We got over fifty sets! Bruce and Ruth Hawkins have been so generous throughout. Jo Whitehouse, in addition to being our board chair through the very rough time when I was sick, has also been a steadfast and extremely generous benefactor. I'm sure we wouldn't have made it without her. Jeanne Shaw, Kim Johnson, Suzi Soza, and Robin Phillips also all collected many of their friends to become donors through wonderful parties and generous donations themselves. I'm also so grateful for Robin's lake house and all the love she has showered on me, my father, and our family (see her genius film on who the real Shakespeare is). Bob Kiekhefer began as a volunteer and then over the years has become one of our best donors! Andrea Borden and her husband, John Gillespie, were the amazing donors who kicked off the medical center fundraising, and Betty Gardner made it all possible. Of late, I also really want to thank Jake Sargent, who made absolutely instrumental donations for scaling and who is one of the wisest donors I know in using his funds to leverage change. Pat Plude and Steve Kusmer not only saved my life and shared so much love and wisdom with me, but also have been some of our biggest donors with both financial resources and time. Pat's wisdom in helping me figure out how to teach radical listening (she even wrote a dissertation on it), Steve's technology brilliance in thinking about scaling, and both of their expertise in nonprofit management have been foundational. Similarly, Michael Bauman's personal generosity and his sharing of his fundraising expertise have been an incredible blessing. Guy Thomas and Monika Kunert have also been so incredibly generous, even making an office for Health In Harmony in San Francisco. We are also so grateful to Vangelis Giorgiou and his family, whom Etty nicknamed, for good reason, the "gorgeous" family and Febri calls "our angels" for their strategic donations and behind-the-scenes political moves! And there are so many others who I would love to talk about but don't have their explicit permission to do so. You know who you are and you know how grateful we are!

We are allowed to list our foundation and institutional supporters, though, and here are some of the key ones: Packard Foundation (Kai Carter, you are the best program officer ever); Arcus Foundation; National Fish and Wildlife Foundation; Softmatter Fund; Sall Family Foundation; A'lani Kailani Blue Lotus White Star Foundation; Tides Foundation; Disney Conservation Fund; Conservation, Food, and Health; Full Circle Fund; Ford Foundation; Water Hope; Mulago Foundation; Dining for Women; Grantham Foundation for the Protection of the Environment; Seneca Park Zoo Society; Straubel Foundation; Bali Zoo; Bank Mandiri; Yale School of Medicine; and Metropolitan United Methodist Church—Staats Memorial fund. Of note, some of our most significant donors have asked to remain anonymous.

I have also been incredibly blessed to be introduced through Clare Selgin Wolfowitz to the U.S. ambassadors to Indonesia and then to become close friends with them. Ambassador Robert Blake and his incredible wife, Sofia, were so supportive of our work, and then Sofia gave me the gift of introducing me to their replacements, Ambassador Joseph Donovan and Mei Chou Donovan. Mei Chou came to Sukadana multiple times, introduced us to many critical people, including Rachel Malik, the wife of the British ambassador, who has been a tireless supporter and also visited ASRI. And one of the very best gifts both Sofia and Mei Chou gave me was allowing me to stay at the ambassador's residence every time I or many of the other ASRI staff came through Jakarta. This was a *huge* gift to be able to be in a lavish, centrally located spot, and it never hurt our connections to tell people where we were staying!

I also want to thank my father for paying for me to join Entrepreneurs' Organization. I was once asked to speak at one of their conferences, and I was surprised to discover that I had found my people. I have often been called a "social entrepreneur," but I didn't realize that the operative word was *entrepreneur,* not *social,* until I met business entrepreneurs. Like me, they were out-of-the-box thinkers and willing to risk themselves to make the world a better place. I have loved being in my forum, and the network of supporters all over the world, including Indonesia, has been a huge boon to our work. Thanks especially to Victor Nunnemaker, Stephen Soper, Pascal Zuta, Pavan Kochar, Devin Koch, Anna Rembold, Stu Chrisp, Jodi Dharmawan, and Sanjay Bhojwani. You have all taught me so much and been a huge emotional support from folks who really get it.

I was also blessed to be given a full scholarship (raised by the previous attendees) to take a six-week Stanford executive training course. Through this, I have met an incredible crew of business leaders from all over the world. A particular shout-out to Vangelis Giorgiou, Enrico Cabarone, Ricardo Marino, Helio Mesquita, Eduardo Grabowsky, Sergio Mota, Nicolas Sauvage, and Sanjay Sinha. Thank you so much for sharing your connections, resources, and vision! Without them we likely wouldn't be in Brazil.

I want to mention again Donella Meadows. *The Wall Street Journal* in March 2008 published a front-page piece on Dana's predictions that said that despite the fact *The Limits to Growth* was lambasted when it came out, we were pretty much right on track for Meadows's worst predictions. Dana and her retinue of amazing thinkers deeply influenced my path in life. We were so sad that she died way too early of meningitis while I was in medical school.

As for friends, I couldn't possibly list all the people who have been great supports for me, but I do want to point out the three women who were the systemic thinkers and the best of friends in medical school: Alison Norris, Anna Hallemeier, and Margaret Bourdeaux. The four of us were not interested in just being doctors; we wanted also to address the fundamental causes of disease. We formed a study group, and these three other women got me through medical school and have been critical to founding and guiding Health In Harmony.

Other friends including Ann Lockhart, Toni Gorog, Scot Zens, Crissy Robinson, Julia Riseman, Brent and Regina Beidler, Erika Jenssen, Sallie Lang, Jeff Lehman, Andrew Ramer, Pat Plude, Steve Kusmer, Peter Mayland, Anne Blackwood, Sheri Hostetler, Nonette Royo, Gina Brezini, Monica Nirmala, Clementine Wamariya (who is such an inspiration to me—read her mind-blowing book *The Girl Who Smiled Beads* about her remarkable life after the Rwandan genocide), and Rhiya Trivedi have all been key supports during many different periods of time in my life. Rhiya particularly helped me see that there were whole other ways to live one's life, and she never took *no* for an answer—which is just what I needed. Andrew Ramer is my angel of guidance and loving friendship (read all his amazing books!). Thank you for being part of my life. And then there is my beloved goddaughter, Kahayag Gabrielle Royo Fay. Being welcomed as a member of her inspiring and magical family is one of the gifts of my life. Thank you Nonette, Chip, Sandrayati, and Kyra.

My spiritual communities and guides have also been a deep wellspring of love and energy over the years. First Mennonite Church has been my spiritual home since residency, and the whole congregation has been so supportive of me personally and of Health In Harmony. Our pastor, Sheri Hostetler, is one of the most amazing people I have ever known, and I feel blessed to count her as a friend. My Quaker pastor, Stan Thornburg, was also such a wonderful support the first time I returned from Indonesia, and we loved having an ex-biker marry Cam and me in his cowboy boots and leathers. And then I credit Bobbi Aqua's no-nonsense therapy and skilled acupuncture for getting me walking again after the jellyfish sting. Janice Wright was also a wonderful partner in personal and spiritual healing while I was ill. Of late, the Fertile Void women's community has become incredibly dear to me as well—thank you, Erin Selover and Nirali Shah, for leading us in what it means to follow. And then I am so grateful to Ida Resi for showing me how amazing the ocean of spiritual depth is and how the closer you get to it, the less you care which religious path you followed to get there. I feel incredibly honored that she calls me her sister. May I live up to this calling.

And then a special thank-you to Stephanie Stevens for teaching me how much joy there can be in life. I have the deepest gratitude that I get to know you. You always ask me the key questions that improve every aspect of my life—and improved this book! I love your fierce dragon-ness and your protective angel wings. You are the best!

And then here are the names of all the ASRI permanent staff from beginning to the time of writing. They are the global pathfinders for where we all need to go: Drg. (Indonesian dental degree) Hotlin Ompusunggu, Dr. Romi Beginta, Wilfirmus Uwil, Clara Sari, Dr. Frans, Maradona Pasaribu, Paharizal, Etty Rahmawati, Muhammad Yusuf, Rahimatul Wasilah, Lidya Kristiana, Supatma, Mardalena, Nani Utari, Agus Supianto, Dr. Made Parulian Tambunan, Yuni, Abdurrahman Dahlan (Utai), Junaidi, Dr. Julfreser Sinurat, Natalia Evalinda Purba, Usup Hamdani, Eka Setiawan, Dr. Ronald Natawidjaya, Nurmala, Januarius Romi, Yudhi, Mirja, Setiawati, Hendriadi, Hamisah, Dr. Lucy Nofrida Siburian, Dr. Robin Andriyanto, Dr. Fitri Juniarta Hutajulu, Romadoni Anggoro, Rahmi Oktarina, Vera Jusnita Sembiring, Adi Bejo Suwardi, Akib, Lucky Verawati, Hebron Oematan, Heri Iman Santoso, Ika Budi Astuti, Heroples, Purwanto, Imam Syafe'i, Maskur, Junaidi, Dr. Verina Logito, Harri Gunawan, Dr. Lasmida Ruth, Muhammad Chumaidi Rahmatulah, Dr. Nurmilia Afriliani, Efan Juniansyah, Maskur, Ngalim, Syarif Mubarak Alaydrus, Jhonyanus, Asnat Apriana Bengngu, Dr. Vina Christiana, Miftah Zam Achid, Yayat Aryadi, Dr. Jeng Yuliana, Fransiscus Xaverius, Indah Prihatin, Syarifah Aqmalia, Juliansyah, Syarif Faizal, Bagus Maulidi, M. Zulkarnaen, Agus Novianto, Alvi Muldani, Mursidi, Idris, Drg. Marisa Thimang, Dr. Hafidz Alhadi Lukmana, Edi Cahyadi, Muhammad Arif Arianto, Syufra Malina Adesita, Erica Jean Pohnan, Dr. William Timotius Wahono, Maria Martha, Rusmadi, Okto Susanto, Reki Roikhan, Dr. Nur Chandra Bunawan, Nurhayati, Tengku Aulia Utami, Indra Lim, Nurul Ihsan Fawzi, Jackson Helms, Dr. Michelle Marcella Karman, Dr. Alvita Ratnasari, Riduan Setiawan, Jeki Sudiana, Oka Nurlaila, Mahardika Putra Purba, Nur Febriani Wardi, Rina Sayekti, Arie Chandra, Fitri Suryani, Tarjudin, Samsudin, Dr. Chaisari M. Turnip, Dr. Mike Lauda, Dr. Deo Develas, Drg. Inggris Amelia, Drg. Grace Evelyn Pardede, Dr. Fifi Florensia, Dr. Nugroho Sondrio Harsono (BBBR), Berkat Fangatulo Gulo, Brigitta (BBBR), Vini Talenta (BBBR), Dewi Susandi (BBBR), Anius (BBBR), Iis Dahlia, Barrata (BBBR), Zakariyanto, Dr. Siti Fatimah Zahra, Qotrun Izza, Rina Juliana, Dr. Jouito

Yakobson Sinaga, Galuh Pravitasari, Devitson, Fangidae, Yohanes Risky S. Ginting (BBBR), Apriliandi, Irvan, Agustina, Dr. Sri Putu Agung, Paramita Kelakan (BBBR), Drg. Monica Ruth Nirmala, Santa Hutajulu, Indri Amanda, Dr. Maria Puspa Kartika, Drg. Prieska Dinda Astriena, Wahyu Hidayat, and Dr. Nur Chandra Bunawan.

In addition, here are the part-time staff, both current and past. Each year, there are also hundreds of community members who help plant trees during the planting season, but unfortunately, we haven't kept records of all their names. The Forest Guardians: Suharjo, Rajali, Antoni, Samsudin, Ali sadikin, Karim, Iwan Setiawan, Syarifudin, Rosni, M. Riduan, Abdul Rahman, Udin B, Amat andayani, Muslianto, Jamsi, Jono karno, Abdurrahman, Sahmadi, Ardianto, Rajali M., M. Arsyad, Laberman, Hasanudin, Hamiji, Daudek, Pendi, Sahpuri, Nurdin, Hermanus (BBBR), Gabriel (BBBR), Asen (BBBR), Tondan (BBBR), Alan Budi Kusuma (BBBR), Damianus Didi (BBBR), Martinus (BBBR), Leternus (BBBR), Cilin (BBBR), Hasidi, Timotius Torit, Amir Mahmud, Sabidin, Yakobus Mely, Turinadi, Rudiansyah, M. Nasir, Sahmadi, Laberman, M. Arsyad, Samsu Rahmani, Saad Arifin, Matius, Kus, Kirno, Safi'I, Abdullah, Sulaiman, Heri Dona, Wisono, Agus, Jainul Abidin, Muslim, Edi, Juliansyah, Genang, Suandi, Ismail, Juhari, Musa, Sri Malianto, Anwar, Riduan A., Jamri, Darmawan, Mariyansyah, Sardan, Yani, Ismail, Suparman, Wawan, and Budin.

The community health workers: Sutinah, Idyawati, Rita Kurnia Safitri, Jumiah, Albekti, Maria Sri Rahayu, Riza Firdianto, Syarifah Yulianti, Umi Kalsum. Lulu Edi Purnomo, Santa (BBBR), Emilia Sulaika, Fitriwati, Aurelia, Susi Ardianti, Diah Mariani, Farida, Jusni, Siti Jubaidah, Nurlinda Mursi, Widya Wahyuni, Mia Lamiatun, Sapuan Nurhadi, Jumrihatik, Maria, Nurhayati, Eka Sulastri, Hefi maulina, Marlena, and Ratnawati.

And the folks who oversee the seedling nurseries: Heridona, Supardi, Saripudin, Jono, Adiyono (BBBR), and Samsudin (BBBR).

Here are the ASRI board members. I want to give a particular shout-out to Dr. Yeni, who has been the board chair for the last five years (this is the same woman who was one of the early ASRI doctors before going on to do her specialty training in psychiatry). She has navigated many tricky bureaucracy issues with skill, patience, and love. We are so blessed to have her leading the helm. And huge gratitude to Dr. Irene, who helped me from the very, very beginning and has never flagged. Our other board members include Romadoni Anggoro, Haji Firdaus MS, Drg. Hotlin Ompusunggu, Pak Semuel, Pak Wilfirimus, Dr. Made Parulian Tambunan, Dr. Julfreser Sinurat, Drg. Erika Lidia Barus, Pak Ridwan Gunawan, Pak Muslimin, Ibu Farah Diba, Ibu Haretta Djumaan Rivai, and myself.

The staff who have worked with us in Madagascar are Manatosa Ny Hasina Fanilo, Andrinidrainy Vao Florent Michel Tojo, Botovelo Sylvère Victorien, Razafimanisa Lalaina Thomas, Raherivololona Rasoamanandray Evelyne Georgette, Rabelafy Sylvain, Razanantenaina Hoby, Rasoanandrasana Robline, Rafalimanana Rija Harinaivo, Mahaleovoninahitra Franckestenny Dhennys Mack, and Ranjemiarisoa Nelly.

And our newest colleagues in Brazil are: Dr. Érika Pellegrino and Marcelo Salazar.

The Health In Harmony employees from the very beginning are Brita Johnson, Rosevan Vickery, Nicole Simpson, Trina Noonan, Darya Minovi, Amy Krzyzek, Bethany Kois, Kari Malen, Kelsey Hartman, Michelle Bussard, Lisa Micek-Hillerns, Loren Bell, Martini Morris, Safia Jama, Sally Oakes, Sara Helms, Tanya Johnston, Jonathan Jennings, Ashley Emerson, Maggie McDow, Devon Schmidt, Thomas Phillps, Kristen Grauer, Devika Gopal Agge, and Adam Burnett. We've also worked closely with these folks on a contract basis: Scott Jones, Annie Jones, Jocelyn Stokes, Nick Viele (thank you for so wisely guiding every nearly every critical meeting Health In Harmony has ever had!), Noortje Trienekens, Emma Sutton, and Dr. Padini Nirmal.

Also, an enormous thanks to the Health In Harmony board, who when we started were also basically staff and who have all given incredibly generously of their time, treasure, and talent. Each of our board members has played absolutely instrumental roles in the development of the organization, but I would be remiss to not specifically point out Alice Prussin who project managed the construction of the medical center in addition to designing all the lighting. There would be no building without her. Petty Taylor has also done an incredible job connecting us to many high-level key Indonesians who have provided invaluable assistance. Thank you especially, too, to our board chairs: Julia Riseman, Anna Hallemeier, Alison Norris, Maggie McDow, Jo Whitehouse, and Emily Scott. In alphabetical order by first name here are all our board members over the years: Aaliya Hamid, Alice Prussin, Alison Norris, Ann Elizabeth Kurth, Ann Lockhart, Anna Gibb Hallemeier, Anne Byram Blackwood, Antonia Gorog, Art Blundell, Brent Beidler, Catherina Celosse, Christina E. Fitch, Christina Sabater, Clare Selgin Wolfowitz, Connie Gersick, Courtney Bergeron, Courtney Howard, Darin Collins, Dave Birckhead, David R. Dishman, Devon E. Wilson-Hill, Elena Bennett, Emily Scott, Enrico Carbone, Gerald George, Hannah Fairbank, Hanneke Jansen, Jan O'Brien, Jeanne Bergman, Jeff Wyatt, Jo Whitehouse, Julia Riseman, Kathleen White, Kevin Boer, Kim Johnson, Kristin Rinehart-Totten, Laszlo Tamas, Laura Tesch, Lester Licht, Luc Janssens, Maggie McDow, Mark Totten, Melanie Webster, Monica Nirmala, Nancy Angoff, Nancy Drushella, Neil Hollyfield, Nicholas Horton, Peter Glassman, Peter Mayland, Petty Taylor, Preetha Rajaraman, Robert Rohrbaugh, Sheryl Osborne, Steve Kusmer, Thomas P. Duffy, Tim Waters, and Vince O'Hara. Michele Barry, Margaret Bourdeaux, and Rhett Butler have also served on our advisory board.

In addition, I want to thank all the volunteers and students who over the years have become part of our planetary health family. Many of our staff or board members started out as volunteers, so their names may also be listed here. Kari Malen is a particularly notable example who volunteered for three years before we hired her for many more. Patrick Ryan came many, many times, and his help was essential; plus the cookies for the HIH staff were greatly appreciated. Richard Ramer worked intensely on finalizing and revising the medical center design and pulled in an amazing team of engineers to help. Dr. Krista Farey has also come multiple times and is so beloved by the ASRI staff for her wonderful teaching. She is now helping us scale our unusual approach to health care, which is based on love, respect, and a 360 approach, including the environment. Here are all the volunteers: Aaliya Yaqub Hamid, Adam Miller, Adeel Chaundhry, Adetania Pramanik, Aghia Gunawan, Ahvigahyel Ayanna Tate, Aik Yang Ng, Aimee Keyashian, Aimee Thornton, Alan Gianotti, Alcan Sng, Alexander Domingo, Alexandra Ristow, Alia Antoon, Aliendheasja Fawilia, Alison Heimowitz, Alison Lee, Alison Norris, Alla Smith, Alyssa Cowell, Amaala Malik, Amalia van den Tempel, Amanda De La Paz, Amanda Storkson, Amy Dear-Ruel, Amy Hurrell, Amy Kravitz, Amy Vagedes, Ana Maria Jimenez Cecilia, Ana Sofia Amieva-Wang, Andrew MacDonald, Andrew Winterborn, Andrew Young, Ann Tseng, Anna Arroyo, Anna Dill, Anna Guttridge, Anna Kobylinska, Anna Ostrowska, Anne Chipman, Anto Purwanto, Anupriya Ramamoorthy, Ariel Fillmore, Ashley Bullers, Ashti Doobay-Persaud, Asil Ibrahim, Audrey Schuetz, August Wilhelm, Austin Weiss, Avneesh Jay Moghe, Bailey Meyers, Bella Jovita, Ben Renschen, Benjamin Johnson, Bernard Boey, Bethany Kois, Brent Beidler, Brian G. McAdoo, Brooke Cotter, Brooks Walsh, Bryan Watt, Brynne Underwood, Camia Crawford, Cara Laviola, Caroline Castillo, Carolynn Fitterrer, Catharina Indirastuti, Cayley Lanctot, Charles and Rosemarie Carodenuto, Charles Rouse, Charles Wang, Chelsea Call, Chloe Hall, Chris Dimock, Chris Moore, Chris Todd, Christina Fitch, Christina Gomez-Mira, Christopher Woerner, Claire Todd, Clancy Broxton, Clara Ottesen, Colin McDonnell, Collin Smith, Corey Schmidt, Courtney Howard, Courtney Nevitt, Cynthia Frary McNamara, Dame Idossa, Damian Okruciński, Dan McCarthy, Dana Schneider, Daniel Albert, Daniel Chow, Daniel Ebbs, Daniel Saada, Danny

Gerber, Darcy Scott, Darin Collins, Darin McClellan Bell, Dave Adams, David Adler, David Allen, David Karam, David Wilson, David Woodbury, Debby Sarita, Debra Cohen, Deepa Agashe, Denice Tai, Derek Richardson, Dhairyashil Rana, Dhara B. Patel, Diane Dakin, Diego Lopez de Castilla, Don Sandev Ferdinando, Donald Grant Gaylord, Dr. Yeni Widiandriany Paonganan, Edward Adi Pranoto, Efrina Paramitha, Elisabeth Marr, Elizabeth Bast, Elizabeth Valitchka, Elyssa Berg, Emma Dean, Emma Murter, Emma Williamson, Erica Pohnan, Erika Jenssen, Esme Cullen, Eunice Martins, Faizan Arshad, Farah Kherun Ola Van 't Land, Federico Gaspar Caro, Felicia Gunawan, Felona Gunawan, Flavia De Souza, Flavia Nobay, Gabriel Chua, Gabriel Mayland, George Hulley, Gordon Wheat, Greg Neidiger, Halard L. Lescinsky, Hannah Huusom, Hayley Blackburn, Heather Danzer, Helen Brunt, Heng Jin Kiat Noel, Herfina Nababan, Holly Eaton, Holly, Jasper, and Adrian Scheider-Feinberg, Ilana Richman, Ingrid Amelia, Irine Vodkin, Jacobsen Lockhart, Jacquelyn Wallace, James G. Wallace, James Heintz, Jan Malmberg, Jan van 't Land, Jane He, Jane Huff, Jane Lester, Jason Warren, Jasper Feinberg, Javier Campo Linares, Jay Modh, Jeanne Noble, Jeff Wyatt, Jeffrey Baldetti, Jeffrey Lane, Jennifer Bell, Jennifer Blair, Jeremy Pivor, Jeremy Sussman, Jesse Turner, Jessica Crawford, Jessie Kittle, Jia Yu Choy, Jim Young, Joan Cockerill, Joanna Mandell, Joe Kallmeyer, John Daniel Ballew, John Gaudet, John Ingram, John Kugler, John Parker Evans, Jonathan Dean, Jonathan Krisetya, Jonathan Westphal Stennicke, Jonty Dean, Josephine Rosegard, Joshua Fleming, Joshua M. Krantz, Judith Bliss, Julia Goar, Julia Kiehlbauch, Julia Lubsen, Julia Paltseva, Justin Mitchell, Kai Swenson, Karen E. Dahl, Karen Ruby Brown, Kari Malen, Karin Gunther, Karl Dietrich, Karl Grunseich, Karly Bishop, Katherine Homes, Kathleen White, Katie Camarata, Kelly Murphy, Kenneth Morford, Kenneth Ryan Petersen, Keva DeKay, Kimberly Herrmann, Krista Farey, Lady Anjani, Lamia Haque, Latha Swamy, Laurel Mayland, Lauren Herbert, Lauren Tobias, Lauren Weber, Lauri Young, Leila Srour, Leslie Gordon, Liu Zhang, Lorcan McKeown, Loren Bell, Lori Chow, Lucia Amieva-Wang, Lucy Wai Yee Loong, Mae Hweei Ooi, Maggie Chin, Manu Uberoi, Marcus Johansen, Margot J. Apothaker, Maria Holst, Marie McCready, Marin Kheng, Marwa Shoeb, Mary Pope, Matt Roi, Megan Helf, Meghana Gadgil, Mei Elansary, Melinda Turner, Miakoda Plude, Michael Mancuso, Michael Turken, Michele Acker, Michelle Moore, Mike Burke, Mindy Clarke, Miroslava Ziolo, Mitchell Edgar, Mochorei Keseolei, Modesta Era Emilda, Molly Nissen, Muram Ibrahim, Nancy Ewen Wang, Nancy Wildman, Natalia Simanjuntak, Natasha Kristina, Nathalia Murillo Rengifo, Nathanael Wilkins, Nathaniel O'Leary, Neha Gupta, Neil Hollyfield, Nicholas Coombs, Nicholas Salazar, Nick Aloisio,

Nikkole Cupp, Nina Finley, Noah Peart, Parimal Deodhar, Patricia and Stuart Billette, Patrick Ryan, Patty Wang, Paul Kramer, Peter Mayland, Putri Timur, Quinn Spinler, Rafael Martinez, Rafika Nutriawani, Rashele Yarborough, Rebecca Sananes, Regina Beidler, Rhidaya Trivedi, Rima Ayusinta, Rob McGlynn, Robert M. Kieckhefer, Robert Mair, Roberto Cipriano, Roman Ki Van 't Land, Rosemarie Carodenuto, Ross Gordon, Ruchit Shah, Ruri Fitriyanti, Ryan Perumpail, Sachiko Oshima, Sallie Lang, Samantha Kaplan, Sandra Kik, Sara Barmettler, Sara Kamran, Sara Whadford, Sarah Condon, Sarah Walpole, Sarun Charumilind, Sauda Bholat, Shannon Donahue, Sharon Santoso, Shaun Cole, Shayma Shamo, Shelina Musaji, Shervin Shahsavari, Signe Rommenholler Petersen, Sisca Wiguno, Sofia Nielsen, Sofie Buhrkall, Sophia Ferretti, Stefan Wheat, Stella Lesmana, Stephanie Foe, Stephanie Gee, Sujana Muttu, Suwitra Wongwaree, Suzanne and Nathan Atteberry, Tamar Gisis, and Tazrin.

It is essentially impossible for me not to have forgotten folks, so please forgive me if your names are not listed. Please just know how incredibly grateful we are nevertheless.

About the Author

Kinari Webb, M.D., is the founder of Health In Harmony—an international nonprofit dedicated to reversing global heating, understanding that rainforests are essential to the survival of humanity—and a cofounder of Alam Sehat Lestari (ASRI). Dr. Webb graduated from Yale School of Medicine with honors and currently splits her time among Indonesia, international site assessments, and the San Francisco Bay Area.

guardiansofthetrees.org